BUILDING SENSOR
NETWORKS
From Design to Applications

Devices, Circuits, and Systems

Series Editor

Krzysztof Iniewski

CMOS Emerging Technologies Inc., Vancouver, British Columbia, Canada

FORTHCOMING TITLES:

Nanopatterning and Nanoscale Devices for Biological Applications
Krzysztof Iniewski and Seila Selimovic

Nanoplasmonics: Advanced Device Applications
James W. M. Chon and Krzysztof Iniewski

Nanoscale Semiconductor Memories: Technology and Applications
Santosh K. Kurinec and Krzysztof Iniewski

Radio Frequency Integrated Circuit Design
Sebastian Magierowski

Smart Grids: Design, Strategies, and Processes
David Bakken and Krzysztof Iniewski

Soft Errors: From Particles to Circuits
Jean-Luc Autran and Daniela Munteanu

Technologies for Smart Sensors and Sensor Fusion
Kevin Yallup and Krzysztof Iniewski

VLSI: Circuits for Emerging Applications
Tomasz Wojcicki and Krzysztof Iniewski

BUILDING SENSOR
NETWORKS
From Design to Applications

EDITED BY
Ioanis Nikolaidis
Krzysztof Iniewski

CRC Press
Taylor & Francis Group
Boca Raton   London   New York

CRC Press is an imprint of the
Taylor & Francis Group, an **informa** business

CRC Press
Taylor & Francis Group
6000 Broken Sound Parkway NW, Suite 300
Boca Raton, FL 33487-2742

First issued in paperback 2017

© 2014 by Taylor & Francis Group, LLC
CRC Press is an imprint of Taylor & Francis Group, an Informa business

No claim to original U.S. Government works

Version Date: 20130722

ISBN 13: 978-1-4665-6272-1 (hbk)
ISBN 13: 978-1-138-07328-9 (pbk)

Library of Congress Cataloging-in-Publication Data

Building sensor networks : from design to applications / editors, Ioanis Nikolaidis,
    Krzysztof Iniewski.
        pages cm. -- (Devices, circuits, and systems ; 17)
    Includes bibliographical references and index.
    ISBN 978-1-4665-6272-1 (hardback)
    1. Sensor networks--Design and construction. 2. Wireless communication
systems--Design and construction. I. Nikolaidis, Ioanis, editor of compilation. II.
Iniewski, Krzysztof, 1960- editor of compilation.

TK7872.D48B83 2013
681'.2--dc23   2013025842

Visit the Taylor & Francis Web site at
http://www.taylorandfrancis.com

and the CRC Press Web site at
http://www.crcpress.com

Contents

SECTION I   Design Practices

SECTION II   Networking Protocols

SECTION III   Application Experiences

About the Editors

Ioanis Nikolaidis received his B.Sc from the University of Patras, Greece, in 1989, and subsequently his M.Sc. and Ph.D. from Georgia Tech, USA, in 1991 and 1994, respectively. He worked as a research scientist for ECRC GmbH (1994–1996) and joined the University of Alberta in 1997 at the rank of Assistant Professor, where he is now (since 2008) at the rank of full Professor. He has supervised, or is currently supervising, a total of 13 Ph.D. and 11 M.Sc. students. He has published 90 papers in refereed journals and conferences, and 4 book chapters. His research interests are in the general area of computer network protocol modeling and simulation, network protocol performance, and wireless sensor network architectures and applications. He is the co-recipient of the best paper award of CNSR 2011 and co-recipient of the University of Alberta Teaching Unit Award as part of the the SmartCondo teaching team. He holds an Adjunct Professor appointment with the department of Occupational Therapy at the University of Alberta. He was Area Editor for Computer Networks, Elsevier (2000–2010), and is Editor (1999–) and has been Editor-in-Chief (2007–2009) for the IEEE Network magazine. He has cochaired CNSR 2011, and ADHOC-NOW 2004 & 2010. He serves as a steering committee member of WLN (annual workshop co-located with LCN) and of ADHOC-NOW. He served as an NSF Panel Member in 2010 and he belongs to the MITACS College of Reviewers (2007–). He has served as technical program committee member and reviewer for numerous conferences and journals, as well as for the following funding agencies: NSERC, NSF, Ontario Centers of Excellence, FWF/START (Austria), NTU/IntelliSys (Singapore), and NWO (Holland). He is a member of IEEE and a lifetime member of ACM.

Krzysztof (Kris) Iniewski is the managing director of research and development at Redlen Technologies Inc., a start-up company in Vancouver, Canada. Redlen's revolutionary production process for advanced semiconductor materials enables a new generation of more accurate, all-digital, radiation-based imaging solutions. Dr. Iniewski is also a president of CMOS Emerging Technologies Research (www.cmosetr.com), an organization of high-tech events covering communications, microsystems, optoelectronics, and sensors. In his career, Dr. Iniewski has held numerous faculty and management positions at the University of Toronto, the University of Alberta, SFU, and PMC-Sierra Inc. He has published more than 100 research papers in international journals and conferences. He holds eighteen international patents granted in the United States, Canada, France, Germany, and Japan. He is a frequent invited speaker and has consulted for multiple organizations internationally. He has written and edited several books for IEEE Press, Wiley, CRC Press, McGraw-Hill, Artech House, Cambridge University Press, and Springer. His personal goal is to contribute to healthy living and sustainability through innovative engineering solutions. In his leisure time, Dr. Iniewski can be found hiking, sailing, skiing, or biking in beautiful British Columbia. He can be reached at kris.iniewski@gmail.com.

List of Contributors

Cesare Alippi
Dipartimento di Elettronica ed
   informazione
Politecnico di Milano
Milan, Italy

Masoud Ardakani
Department of Electrical & Computer
   Engineering
University of Alberta
Edmonton, Canada

Giacomo Boracchi
Dipartimento di Elettronica ed
   informazione
Politecnico di Milano
Milan, Italy

Swastik Brahma
Department of Electrical Engineering
   & Computer Science
Syracuse University
Syracuse, New York

Stephane Bressan
School of Computing
National University of Singapore
Singapore, Singapore

Mainak Chatterjee
Department of Electrical Engineering
   & Computer Science
University of Central Florida
Orlando, Florida

Ting-Shuo Chen
Department of Electrical Engineering
National Taiwan University of Science
   and Technology
Taipei, Taiwan

Chi-Tsun Cheng
Department of Electronic and
   Information Engineering
The Hong Kong Polytechnic University
Hung Hom, Kowloon, Hong Kong

Saverio De Vito
ENEA: Ente Nuove Tecnologie,
   Energia e Ambiente
Centro Ricerche Portici
Portici Naples, Italy

Hassan Ghasemzadeh
Computer Science Department
University of California at Los Angeles
Los Angeles, California

Ann Gordon-Ross
Department of Electrical and
   Computer Engineering
University of Florida
Gainesville, Florida

Jeff Hiner
Department of Electrical and
   Computer Engineering
University of Arizona
Tucson, Arizona

Roozbeh Jafari
Electrical Engineering Department
University of Texas at Dallas
Dallas, Texas

Hoyoung Jeung
SAP Research
Brisbane, Australia

Thomas Kister
School of Computing
National University of Singapore
Singapore, Singapore

Chung-Hsien Kuo
Department of Electrical Engineering
National Taiwan University of Science
    and Technology
Taipei, Taiwan

Francis C. M. Lau
Department of Electronic and
    Information Engineering
The Hong Kong Polytechnic University
Hung Hom, Kowloon, Hong Kong

Susan Lysecky
Department of Electrical and
    Computer Engineering
University of Arizona
Tucson, Arizona

Roman Lysecky
Department of Electrical and
    Computer Engineering
University of Arizona
Tucson, Arizona

Baljeet Malhotra
SAP Research
Singapore, Singapore

Richard Mietz
Institute of Computer Engineering
Universität zu Lübeck
Lübeck, Germany

Sahar Movaghati
Department of Electrical & Computer
    Engineering
University of Alberta
Edmonton, Canada

Arslan Munir
Department of Electrical and
    Computer Engineering
University of Florida
Gainesville, Florida

Kay Römer
Institute of Computer Engineering
Universität zu Lübeck
Lübeck, Germany

Manuel Roveri
Dipartimento di Elettronica ed
    informazione
Politecnico di Milano
Milan, Italy

Shamik Sengupta
John Jay College of Criminal Justice
City University of New York
New York, New York

Ashish Shenoy
Department of Electrical and
    Computer Engineering
University of Arizona
Tucson, Arizona

Wei Su
Communications-Electronics RD&E
    Center
US Army RDECOM
Aberdeen Proving Ground, Maryland

Kian-Lee Tan
School of Computing
National University of Singapore
Singapore, Singapore

Chi K. Tse
Department of Electronic and
    Information Engineering
The Hong Kong Polytechnic University
Hung Hom, Kowloon, Hong Kong

Alan Wilson
Maritime Platforms Division
DSTO Defence Science and
    Technology Organisation
Victoria, Australia

Jefferson L. Xu
Cirrus Logic
Austin, Texas

Mengchu Zhou
Department of Electrical & Computer
    Engineering
New Jersey Institute of Technology
Newark, New Jersey

Introduction

The building of Wireless Sensor Networks (WSNs) continues to capture the imagination of a large number of researchers, scientists, and engineers as well as hobbyists and practitioners. Clearly, developing a system relying on nodes of sometimes-unremarkable capabilities is a unique challenge. In a sense, WSNs are a form of "retro" computing that strikes a nostalgic chord of "trying to do more with less" that runs against the zeitgeist of abundant computing and storage resources. The appeal of WSNs is further enhanced by a conspiracy of several interested parties that are rarely seen together. For starters, scientists find in WSNs the potential to utilize them to perform better science by deploying nodes abundantly and closer to the area of phenomena that they want to observe or control. Hobbyists and practitioners see in WSNs the potential of trying out novel ideas and practical gadgets without having to take on a mortgage to afford the necessary equipment. Computer scientists see in WSNs the embodiment of large-scale vertices-and-edges abstractions coupled with elementary computing, rekindling their interest in truly distributed computing. Electrical engineers see in WSNs the potential for a signal-processing and -control coordination platform of exceptional flexibility. The list of WSN fascination goes on and on. In other words, their attraction lies in how you build them as well as (once built) what you can do with them.

Nevertheless, stepping back and looking at the big picture, one is bound to remark that, for all the interest WSNs have created over the past decade (if not more), there are few examples to show that they are truly delivering on the promise and anticipation that has fueled their popularity as an area of study. Could it be that we are missing a key ingredient? Should we wait for a "killer app" or for the market to determine if WSNs are a "winner" technology or not? While market dynamics or a grassroots need for a particular application could determine whether WSNs enter the mainstream or die off by the wayside, it is fruitful to study the facets that drive WSN development conceptually as well as practically. This book attempts to punctuate those facets that, we believe, have been relatively underserved in the existing literature. For this reason, the chapters in this book thematically deviate from the classical viewpoint that sees WSNs as, in effect, two problems: routing and energy efficiency. While both are important considerations, we reduce the emphasis on them because routing has been studied to exhaustion, and energy efficiency has become a "reflex" consideration to the point that it has to be assumed that each and every publication written on WSNs is read from this viewpoint. Could we then move the focus tangentially, but in a revealing way, such that we could unearth the facets that, sometimes obliquely, are more important than squeezing the last nano-Joule of energy out of a node?

In this book, we offer a narrative that attempts to stitch together the path from conceptual development of applications, on one end, to actual complete application at the other end. We forewarn the reader that we will fail on the side of the complete application. This is not to say that the applications do not exist (quite the contrary), but they are applications that "fail" the purist definition of an ideal WSN in some respect. In some cases, the nodes are wired; in other cases, the nodes are, maybe, "expensive." Yet, as we will see in the applications chapters at the end of this book, the finished systems collectively cover the desired features we would like to see out of a WSN. We hope that by pointing to them in the last section of this book, we are presenting the "lessons learned" from systems that have accumulated a history of nontrivial efforts at development and deployment.

In the first section of the book, entitled "Design Practices," we explore alternative ways to approach the overall task of (a) coming up with a suitable WSN solution to an application and (b) assisting such a development in a manner that, while it is informed by the application, need not be tied to a particular application. The approach advocated in Chapter 1 (Dynamic Profiling and Optimization Methodologies for Sensor Networks) is one whereby an existing system is characterized and tuned by a dynamic profiling and optimization process. High-level metrics important to the user are assigned an importance that is reflected in the objective functions. Still, preserving application independence, Chapter 2 (Stochastic Inference in Wireless Sensor Networks) asserts that a large number of WSN applications are in essence forms of stochastic inference as it is applied to vectors of measurements acquired by those nodes. Hence, the authors propose a factor-graph approach to modeling and solving inference problems, with the factor graphs captured by the processing and communication of the nodes. In a more traditional approach to potent abstractions expressing a large application domain, Chapter 3 (Implementation of Wireless Sensor Network Systems with PN-WSNA Approaches) suggests the use of Petri net models with suitable extensions while acknowledging the legitimate scalability concerns that point to the need for generating simplified models. Concluding Section 1 is Chapter 4 (Real-Time Search in the Sensor Internet), which, while at first appearing to narrow the application domain, points to the inescapable reality that many applications of WSNs are going to interact with the traditional Internet infrastructure, primarily providing search-like services. The chapter presents both a survey of existing approaches as well as a possible semantic search engine for the sensor Internet.

The second section of this book, "Networking Protocols," illustrates the impact of the intermediaries—the "glue" of putting applications together. Network protocols are geared to establishing good performance with respect to certain metrics. In trying to pull together some of the multiple concerns of handling traffic, Chapter 5 (Traffic Management in Wireless Sensor Networks) points to distributed congestion-control algorithms to adaptively address the fair and efficient handling of traffic. Another way to constrain the protocols is by appreciating that the traffic and interaction needs of nodes are directly influenced by the application. Specializing the protocols to body-area (wearable) networks, Chapter 6 (Decision-Tree Construction for Event Classification in Distributed Wearable Computers) turns the tables around by putting first the application demands (action recognition) and then localizing the traffic and interactions by grouping nodes together in a manner such that each action recognition can be performed efficiently using decision-tree classifiers. The more "traditionalist" approach to separating a network into clusters (primarily to locally regulate transmissions and save energy) is described in Chapter 7 (A Network Structure for Delay-Aware Applications in Sensor Networks) that, as its title suggests, tries to counter the common deficiency of clustering, i.e., the delay added to the data-collection tasks. Section 2 concludes with Chapter 8 (Distributed Modulation Classification in the Context of Wireless Sensor Networks), which digs deeper into the layers in WSNs and expresses the view at the physical layer, where the coexistence of a WSN with other systems on a frequency band necessitates the classification of the channel condition. The proposed strategy exploits the "strength in numbers" of the WSN nodes to assist them in collectively identifying the presence of primary users in a cognitive radio setting.

The concluding section of the book, "Application Experiences," starts with the challenges facing WSN development in usually unfriendly environments to humans, such as in the presence of toxic and dangerous chemical agents. Chapter 9 (Challenges in Wireless Chemical Sensor Networks) points to the significant efforts required in WSNs tuned to chemical application, with special attention to calibration and sensing stability, energy efficiency, and humanly accessible visualizations of the data collected. Delving further into the design practice, Chapter 10 (Low-Power, Extensive Sensor Networks from the Wired Perspective) gives us a glimpse into platforms used by the Defence Science and Technology Organisation (DSTO) of Australia, which takes us on a journey all the way from the requirements of the design to the flow charts of the code driving the low-power platform design. If it does not fit the definition of being an "exotic" application, it abundantly illustrates the need for a meticulous process to generate production-grade working systems. In a somewhat different vein, we are given food for thought in Chapter 11 (Maritime Data Management and Analytics: A Survey of Solutions Based on Automatic Identification System) about how applications of low-bandwidth and relatively sparse communication can have, despite its limitations, significant impact when applied to a particular context. The resulting data, once related to locations, speeds, etc., provide a first-class example of how, while the sensors themselves are not "stealing the show," the data management based on the data they produce has implications on global trade and security. In other words, a limited sensor system does not imply that the applications it supports are insignificant. Keeping with the maritime theme, the final chapter of this book (Chapter 12, Above and Below the Ocean Surface: A WSN Framework for Monitoring the Great Barrier Reef) demonstrates the impact that WSNs can have in our affairs with this planet with respect to maritime ecosystems and how techniques of WSN development and integration have been combined to build such a network.

While we are certainly not suggesting that this book will "solve your problems" of putting together a WSN, we hope that the collection of chapters presented here will illuminate and influence your perspective and priorities when thinking about how to tame the complexity of designing a WSN application. We have deliberately moved the emphasis to be in line with the need to build applications, and we also wanted to present examples that illustrate what applications of WSNs could look like and what constraints the applications bring to the table. We hope that our objective has been accomplished and that you will get as much enjoyment out of going through the pages of this book as we had in putting it together.

Kris Iniewski and Ioanis Nikolaidis
January 2013

Section I

Design Practices

1 Dynamic Profiling and Optimization Methodologies for Sensor Networks

*Ann Gordon-Ross, Arslan Munir, Susan Lysecky,
Roman Lysecky, Ashish Shenoy, and Jeff Hiner*

CONTENTS

1.1   OVERVIEW

The commercialization of sensor-based platforms is facilitating the realization of ubiquitous computing environments previously existing only in science fiction. Numerous platforms have emerged, with endless application possibilities. However, as these platforms continue to evolve, they are becoming increasingly complex, even described as requiring "2.5 PhDs" to design and implement a given application. While many tools exist that robustly model the underlying hardware and communication channels, application-level information remains difficult and tedious to capture. Currently, application experts are left to specify application behavior using input files, mathematical models, or synthetic data generation. A few efforts have appeared that attempt real-time sensor-based platform monitoring; however, these approaches can incur significant overhead, and thus reduce the lifetime of the systems. Currently, an accurate and efficient method to capture external application-specific stimuli remains elusive.

Additionally, much of the complexity associated with sensor-based platforms is due to the plethora of parameters that must be considered when implementing an application. Application experts must balance a large amount of competing parameters. While the effects that various parameter configurations have on high-level metrics, such as lifetime, have been well documented, applying this information remains difficult. To further complicate matters, application experts do not necessarily have programming or engineering expertise, but rather are biologists, teachers, or agriculturists who wish to utilize the sensor-based platform within their given domain. Thus, faced with an overwhelming number of choices, application design may be a daunting task for nonexperts.

In this chapter, we describe methods aimed at alleviating some of the complexities associated with sensor-based system design through the use of computer-aided design techniques. We also present a dynamic profiling and optimization platform (DPOP) capable of observing application-level behavior and dynamically tuning the underlying platform accordingly. Whereas several research groups are focusing on the specification of sensor networks using different programming paradigms [1–23], our work focuses on optimizing the underlying platform.

1.2   INTRODUCTION

Numerous sensor-based platforms have appeared, enabling a wide range of application possibilities. With each application scenario, developers have a unique set of *application requirements,* such as lifetime, responsiveness, reliability, or throughput, that must be met. For example, a disaster-response application requires high responsiveness and reliability to survey damage or detect survivors, but may only require a lifetime of days or weeks. Conversely, an automated vineyard irrigation system would have a longer lifetime requirement, as it is intended to operate on the order of years.

To achieve various application goals, commercial-off-the-shelf (COTS) sensor nodes possess *tunable/configurable node-level parameters,* such as voltage levels, processor mode, or configurable baud rates [24], and *protocol-level design choices,* such as power cycling to sensing units [25] or data aggregation and filtering [26]. While the effects that various parameter configurations have on high-level design metrics have been well documented, balancing these numerous competing metrics remains challenging. To further complicate matters, predicting application behavior is extremely difficult at design time. Tuning the underlying platform to inaccurate application behavior estimates can yield either suboptimal results or negatively impact the resulting system. Currently, application experts are left to specify application behavior via an input file [27], a mathematical model [28], or through synthetic data generation [29]. While a few real-time tools have appeared [30], they are not designed for dynamic profiling and incur significant overhead.

To alleviate some of the complexity associated with application-specific tuning of sensor-based systems, we have begun to develop a dynamic profiling and optimization (DPOP) framework. Dynamic profiling and optimization not only reduces designer effort, but an automated environment increases accessibility to application experts (e.g., biologists, educators, agriculturalists, etc.) by abstracting much of the underlying platform-specific knowledge. Dynamic profiling enables an accurate view of the application behavior while the system is immersed in the intended environment, eliminating the guesswork of trying to create a "good" benchmark. Furthermore, by profiling applications dynamically, we can monitor how the application responds to changes in environmental conditions or changes in the underlying platform (e.g., a node is disabled), opening opportunities to dynamically reoptimize and update the underlying platform accordingly.

However, such dynamic optimization presents two major challenges: (a) collecting accurate profiling results at runtime while balancing the overhead incurred and (b) developing efficient dynamic optimization algorithms/heuristics to tune the configurable parameters to meet application requirements based on the application profile. For the first challenge, dynamic profiling—an accurate and robust method to capture external application-specific stimuli—remains elusive. While many dynamic profiling techniques exist in the embedded-systems and architecture domain, these techniques are highly specific to their intended system and thus are quite low level. To the best of our knowledge, there exists no previous work in dynamically profiling for sensor networks. Furthermore, the distributed nature of sensor-based networks complicates adoption of previously developed profiling methods to sensor-based networks. One of the major challenges of dynamically profiling sensor-based platforms is accurately capturing application behavior without incurring significant overhead or significantly altering system behavior.

In the second challenge, dynamic optimization algorithms determine the appropriate values for tunable parameters to meet application requirements. However, given the large design space, determining appropriate parameter values (operating state) is challenging. Typically, sensor-node vendors assign an initial generic tunable parameter setting; however, no one tunable parameter setting is appropriate for all applications. *Parameter optimization* is the process of assigning appropriate (optimal or near optimal) tunable parameter values dynamically during runtime in response to changing environmental stimuli observed via the dynamic profiling information to meet user-specified goals. There exists much previous work in dynamic optimizations [31–36], but most of this work targets the processor or memory (cache) in computer systems. There exists little previous work on WSN dynamic optimization, which presents a more challenging endeavor, given a unique design space, resource constraints, and platform particulars as well as a sensor network operating environment.

In this chapter, we present several profiling methods for dynamically monitoring sensor-based platforms and analyze the traffic/energy/code impacts for two prototyped sensor-based systems, and we investigate an appropriate approach to implementing the parameter optimization given the dynamic profiling data already collected from the platform. Next, we propose a lightweight dynamic optimization methodology to quickly determine the best operating state with respect to the dynamic profiling data and user specifications. Using an evaluation of parameter-value effects on the operating state, we propose a methodology to determine appropriate initial tunable parameter values, which determines a high-quality operating state in *one shot*. To improve upon the one-shot operating state, we propose an intelligent exploration ordering of tunable parameter values and an exploration arrangement of tunable parameters, and leverage this information to quickly explore the design space using an online greedy algorithm. Whereas many research groups are focusing on the specification of sensor networks using different programming paradigms [1–23], our work focuses on optimizing the underlying platform.

1.3   RELATED WORK

1.3.1   DYNAMIC PROFILING

Dynamic optimization relies upon accurate profiling results collected at runtime. *Currently*, an accurate and robust method to capture external application-specific stimuli remains elusive. While many dynamic profiling techniques exist, these techniques are highly specific to their intended system and thus are quite low level. For example, working-set analysis [37] monitors the current set of executing instructions to determine changes in system execution. Kaxiras, Hu, and Martonosi [38] determined changes in cache requirements using counters embedded within the cache structure, while other methods simply observe current idle periods [39]. Idle-period observation is a generalized, high-level mechanism to profile a system, but when it is applied to sensor networks, little information on overall system behavior can be inferred.

The distributed nature of sensor-based networks complicates the adoption of previously developed profiling methods to such networks. One of the major challenges to the dynamic profiling of sensor-based platforms involves the accurate capture of application behavior without incurring significant overhead or significantly altering system behavior. In many simulation frameworks, application experts are left to specify application behavior via an input file [27], a mathematical model [28], or through synthetic data generation [29]. A proposed alternative solution involves the use of sensor network emulators that can enable control of particular sensor nodes, providing controllability and repeatability for testing, evaluating, and comparing sensor networks [30]. However, while the proposed emulation and profiling framework is ideally suited to developing and benchmarking sensor networks in a reliable and repeatable manner, its use in deployed systems would incur significant overhead and is not directly applicable. Additionally, application-layer tools exist that provide the required server software services. Specifically of interest is the support of real-time monitoring of a deployed sensor network that provides visualization and data collection capabilities. However, while the proposed real-time monitoring techniques are applicable to a system in the field, the overhead of these services can be significant [9].

The battery remains a dominant constraint in many sensor network applications, resulting in the proposal of simulation frameworks that integrate power estimation. In one power-profiling technique, current draw is premeasured for a variety of CPU modes as well as the sensor board and EEPROM [40]. These values are integrated into to an event-driven simulator for TinyOS applications [14] to determine how much time each subcomponent spends in a particular mode of operation, thereby calculating power consumption of individual nodes. Eriksson et al. [41] similarly have strived to estimate the network-level power consumption through a combination of the COOJA and MSPSim simulator.

In addition to node-level power estimation, which similarly combines time spent in different operating modes with premeasured consumption of these components, a network simulator can be integrated into the framework to account for communication between nodes based on external emulation of the radio chip, sensor boards, and flash memory. Simulation frameworks can also provide information pertaining to low-level hardware and network parameters. For example, the TOSSIM simulator [14] can track statistics such as packet loss, CRC failure rates, as well as the length of send queues, while Avrora [42] monitors hardware interrupts, I/O registers, and memory usage. The ATEMU framework [27] additionally provides application experts with insight into the number of backoffs performed after transmission collisions. While Avrora and ATEMU are cycle-accurate instruction-level simulators, TOSSIM is an interrupt-level discrete-event simulator. Furthermore, the addition of energy models to these simulators, like AEON to Avrora and PowerTOSSIM to TOSSIM, enables estimation of power consumption [43], as mentioned previously.

XRM's reactive-modules simulator employs statistical methods and high-level-state machine modules to simulate a sensor network [44]. While providing a fast simulation approach, these statistical approaches often have insufficient capability to reproduce unique or aberrant sensor events, and it is often difficult or infeasible to produce a statistical model that captures the behavior of a real application. CENSE is a different approach that combines simulation methods with hardware emulation of physical sensor nodes to provide accurate low-level simulation capabilities [45]. This combined simulation/emulation approach can provide very accurate measurements. For example, by coupling the simulation approach with power-measurement hardware, an accurate analysis of the sensor network's power consumption can be determined. However, such emulation approaches require specialized hardware for each emulated physical sensor node, and they are difficult to scale to large sensor network simulations. In addition, because the emulation hardware is specific to the nodes being tested, the model is difficult to generalize or adapt to other sensor-node hardware.

While there is no doubt that simulation frameworks are an essential part of a developer's tool kit to evaluate and test prototype designs, developers must still make assumptions or predictions about the deployment environment, which can lead to inaccurate results. Furthermore, as Handziski et al. [46] denoted, the lack of a wide range of protocols and the lack of customization of models also add uncertainty to simulation results. Thus, we have attempted to avoid many of these issues simply by performing profiling while the sensor network is deployed in its intended environment, which additionally opens the opportunity to dynamically optimize as the system stimuli or underlying platform change over time.

1.3.2 DYNAMIC OPTIMIZATIONS

There exists much research in the area of dynamic optimizations [31–36]; however, most previous work has focused on the processor or memory (cache) in computer systems. Whereas previous work can provide valuable insights into WSN dynamic optimizations, these works are not directly applicable due to a WSN's unique design space, energy constraints, and operating environment.

In the WSN domain, Sridharan and Lysecky [47] implemented a dynamic profiling methodology to obtain environmental stimuli by instrumenting the underlying sensor-based platform while operating in its intended environment. However, a methodology to leverage these profiling statistics to optimize the underlying platform is not presented. Shenoy et al. [48] expanded these profiling methods, implementing a variety of dynamic profiling methods on the IRIS sensor network platforms and measuring the associated network traffic and energy. However, this work similarly focused on implementation and customization of the profiling methods and did not explore dynamic optimizations.

To bridge the gap between profiling and optimization, Munir and Gordon-Ross [49] proposed a Markov decision process (MDP)–based methodology as a first step toward WSN dynamic optimizations. However, the MDP-based method required prohibitively large computational resources for larger design spaces and would ideally run on a base station node with more computing resources to carry out the optimal operating-state determination process. The excessive computational requirements inhibited the methodology's implementation on resource-constrained sensor nodes, limiting the nodes' ability to autonomously determine the optimized operating state.

Wang et al. [50] approaches the optimization problem through a distributed energy optimization method for target-tracking applications. The energy management mechanism consists of an optimal sensing scheme that leverages dynamic awakening of sensor nodes. The dynamic awakening scheme awoke the group of sensor nodes located in the target's vicinity for reporting the sensed data. The results verified that dynamic awakening combined with optimal sensor-node selection enhanced the WSN energy efficiency. Liu, Zhang, and Ma [51] also proposed a dynamic-node collaboration scheme for mobile target tracking in wireless camera sensor networks. (Wireless camera sensor networks can provide much more accurate information in target-tracking applications as compared to the traditional sensor networks.) The proposed scheme comprised two components: a cluster head election scheme during the tracking process and an optimization algorithm to select an optimal subset of camera sensors as the cluster members for cooperative estimation of the target's location. Khanna, Liu, and Chen [52] proposed a reduced-complexity genetic algorithm for secure and dynamic deployment of resource-constrained multihop WSNs. The genetic algorithm adaptively configured optimal position and security attributes by dynamically monitoring network traffic, packet integrity, and battery usage.

At a lower level, several papers have explored dynamic voltage and frequency scaling (DVFS) for reduced energy consumption in WSNs. Min, Furrer, and Chandrakasan [53] demonstrated that dynamic processor voltage scaling reduced energy consumption by 60%. Similarly, Yuan and Qu [54] studied a DVFS system that used additional transmitted data packet information to select appropriate processor voltage and frequency values. Although DVFS provides a mechanism for dynamic optimizations, considering additional sensor-node tunable parameters increases the design space and the sensor node's ability to better meet application requirements. To the best of our knowledge, our work is the first to explore an extensive sensor-node design space.

Lysecky and Vahid [55] proposed simulated annealing (SA)–based automated application-specific tuning of parameterized sensor-based embedded systems and found that automated tuning can better meet application requirements by 40% on average as compared to a static configuration of tunable parameters. Verma [56] considered SA-based and particle swarm optimization (PSO) methods for automated application-specific tuning and observed that an SA-based method performed better than PSO because PSO often quickly converged to local minima. Exhaustive search algorithms have been used in the literature for performance analysis and comparison with heuristic algorithms. Meier et al. [57] proposed an exhaustive search-based scheme called NoSE (Neighbor Search and link Estimation) for neighbor search, link assessment, and energy consumption minimization.

Even though there exists some work on optimizations in WSNs [50, 53, 54, 58–61], dynamic optimizations require further research and more work in depth considerations. Specifically, a sensor node's constrained energy and storage resources necessitate lightweight dynamic optimization methodologies for tuning the parameters of a sensor node.

1.4   DPOP ENVIRONMENT

Figure 1.1 illustrates the proposed dynamic profiling and optimization platform. Three main components contribute to the proposed environment: the sensor-based application, the end-user design-metric specification, and the dynamic profiling and optimization (DPOP) module.

The *sensor-based application* is the physical deployment of the application within the intended environment; it consists of sensor nodes, intermediate processing and routing nodes, and actuator nodes, all working together to achieve the desired application functionality. Ultimately, the application developer is interested in high-level system metrics, such as the expected lifetime of a node or sensor network utilizing the given energy reserves (e.g., two AA batteries), the time required to process a single packet, or the time required to process and respond to a sensor. The *end-user design-metric specification* allows an application developer to define which design metrics are important to a particular application and, then, to choose the acceptable or

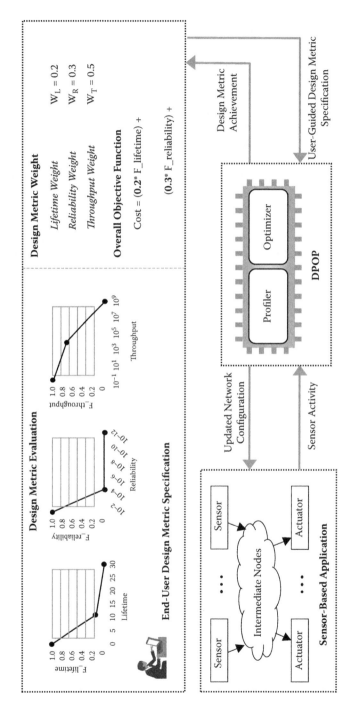

FIGURE 1.1 Dynamic profiling and optimization (DPOP) environment highlighting various tasks, including specifying design metric evaluation equations and assigning design metric weights to indicate the relative importance between each design metric.

unacceptable values of each design metric, thereby providing a method to interpret the resulting system achievement within the context of a given application. First, for each design metric, an application developer correlates a design cost to the raw metric value (e.g., lifetime of 2 months), where a lower design cost corresponds to a superior design-metric value. Figure 1.1 illustrates three design-metric objective functions corresponding to lifetime, reliability, and throughput, for which a graphical interface [55] is utilized to specify the design-metric objective functions as piecewise linear functions. Second, to determine the relative importance of each design metric, the application developer additionally specifies weights for each design metric, as shown in Figure 1.1. The overall objective function—or overall design cost—combines the individual design-metric weights as well as the resulting costs assigned by each design-metric objective function. This overall design cost indicates how well an individual node configuration meets the specified application requirements.

The *dynamic profiling and optimization* (DPOP) node is a separate component—implemented either within the base station node or as a separate sensor node—dedicated to the *profiling* and *optimization* of the underlying sensor-based platform as the platform interacts within the intended environment. The *profiler module* dynamically monitors the application behavior while the sensor-based system is deployed, tracking statistics of interest to the application developer. The *optimizer module* evaluates possible node configurations within the design space to determine which configuration best meets the application requirements. Given the design-metric evaluation specification and dynamic-profile data, the optimizer first utilizes an equation-based estimation methodology that estimates each design metric using both the node configuration and profile data. The optimizer then explores the design space by evaluating each feasible node configuration to determine which node configuration is best suited for a given application (i.e., the configuration yielding the lowest overall design cost). Dynamic optimization of sensor nodes using the DPOP environment can yield up to an 83% improvement in overall design costs compared to a statically optimized node configuration.

1.5   DYNAMIC PROFILING OF APPLICATION BEHAVIOR

1.5.1   DYNAMIC PROFILING METHODOLOGY

Within the DPOP environment, dynamically profiling a sensor-based application requires profiling methods to be incorporated within each node to monitor the execution behavior for individual sensor nodes. Additionally, in order to optimize a sensor-based system, a global view of the entire system is needed. As such, the resulting node-level profile data must eventually be transmitted and analyzed by the system-level profiler module. Numerous profiling strategies can be employed to collect the pertinent application-level information. As highlighted in Figure 1.2, each profiling strategy must consider: (a) what application-level parameters need to be profiled, (b) whom to profile within the network, (c) when to perform profiling, and (d) how to transmit profile information from the individual sensor nodes.

One of the foremost concerns for a profiling strategy is to determine what low-level execution statistics (e.g., sensor sampling rate, packet transmissions, battery

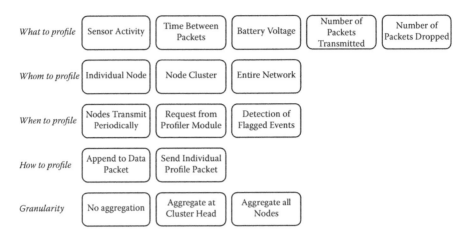

What to profile	Sensor Activity	Time Between Packets	Battery Voltage	Number of Packets Transmitted	Number of Packets Dropped

Whom to profile	Individual Node	Node Cluster	Entire Network

When to profile	Nodes Transmit Periodically	Request from Profiler Module	Detection of Flagged Events

How to profile	Append to Data Packet	Send Individual Profile Packet

Granularity	No aggregation	Aggregate at Cluster Head	Aggregate all Nodes

FIGURE 1.2 Overview of profiling strategies considerations, including *what, whom, when,* and *how* to profile.

charge) need to be profiled within the deployed network. At each sensor node, various low-level execution details must be monitored in order to enable the optimization approach to accurately estimate the various high-level design metrics of interest. For example, consider a sensor node that periodically samples and reports sensor data. The power consumption of the software required to process each sensor event and transmit the packet can be statically measured through physical measurements. The runtime power consumption of the software executing on a node can be estimated as a function of the measured power consumption and the dynamic sensor sampling rate. Overall, determining what low-level metrics to profile within a sensor-based platform is thus related to both the high-level design metrics of interest and the estimation method utilized to evaluate those design metrics. Within our current profiling implementation, both the *sensor sampling rate* and the *time between successive packets* can be profiled for individual sensor nodes.

Profiling individual sensor nodes is not always necessary. Hence, a profiling strategy must consider the granularity at which profiling is performed, including profiling an individual node, a cluster of nodes, or the network as a whole. The granularity at which to profile is affected by both application and network topology. For example, the profiler may want to profile only those intermediate nodes whose job is to forward packets, as these nodes may have higher energy consumption, where optimizing lifetime is of critical importance. Alternatively, the profiler may consider a single sensor node placed in a known area with high activity to determine the minimum sampling rate needed by the application. Profiling the entire network is also possible, in which nodes directly aggregate (or average) the collected profile information as it is forwarded through other nodes' profiler modules. However, profiling at different levels of granularity provides the ability to trade off profiling detail with profiling overhead. Although various profiling granularities are possible, our current profiling implementation only supports profiling of *individual nodes*.

Given the desired profiling information to be collected, the frequency at which profiling is performed directly impacts both the accuracy of the profile data as well as the intrusiveness of the profiling method. On the one hand, profiling can be performed periodically at each node or cluster of nodes. Although the performance and energy overhead of periodic profiling is often easily predicted, the dynamic activity patterns of individual nodes or across the sensor network may be unpredictable, and a periodically collected profile may not accurately describe the current execution behavior. On the other hand, nodes can directly detect all events related to the required profiling information and directly transmit those event occurrences to the profile module as they occur. Such an approach provides the advantage of highly accurate profile information, but it comes at the expense of potential increases in both the code size needed to detect such events and the increased packet transmission overhead due to the occurrence of unpredictable events. An alternative approach to control the collection of profile information is to require the profiler module to explicitly send a profile-request packet. While a packet-transmission overhead is incurred to transmit the profile-request packet, the profiler module can dynamically control how often these requests are sent based on the data collected thus far or on the patterns previously observed. Our current profiling implementation provides support for all three methods of controlling when profiling is performed for individual nodes, specifically: *periodic*, *event-driven*, and *profiler-module directed*.

Finally, the method of transmitting the collected profile data back to the profiler module directly impacts both the network traffic of the sensor network as well as the node's energy consumption, as the radio subsystem must remain active for longer durations to transmit the profile data. Currently, our profiler implementation provides support for either transmitting profile data as *separate profile packets* or appending (i.e., *piggybacking*) the profile data to existing packets already transmitted by the application. Requesting nodes to send separate profiling packets may increase overall network traffic, as each dedicated profile packet must also include the packet header. Alternatively, by piggybacking the profile information onto existing data packets, the profile data can be transmitted without requiring an additional packet header. However, piggybacking profile data onto existing data packets may require individual sensor nodes to store the profile data until the sensor nodes transmit a data packet.

To evaluate the feasibility of the proposed profiling methodologies within the DPOP framework, we implemented these profiling methods on the Crossbow IRIS platform as a set of software functions that can be readily integrated within a sensor application. While incorporating these profiling methods into the target application currently requires designer effort, the required changes do not directly impact the main application functionality. Rather, these methods are inserted within the underlying software infrastructure for packet transmission/reception and sensor interfaces. As many application experts will not need to directly modify this code, various profiling methods can be selected by simply including the required set of software driver source files for the target application.

The resulting profiling methodologies can support all combinations of the previously mentioned *what*, *whom*, *when*, and *how to* profile. However, we currently consider four specific profiling methods:

- PM1: sensor sampling rate and time between successive packets; individual nodes; profiler-module directed; piggybacked

- PM2: sensor sampling rate and time between successive packets; individual nodes; profiler-module directed; separate profile packets
- PM3: sensor sampling rate and time between successive packets; individual nodes; periodic; separate profile packets
- PM4: sensor sampling rate; individual nodes; profiler-module directed; separate profile packets

1.5.2 Experimental Results

We consider two sensor-based applications to evaluate the overheads of the four proposed profiling methods. The first application is a Forest Fire Detection and Propagation Tracking system. During the normal observation mode, individual sensors sample and transmit the surrounding temperate reading every 5 minutes to a base station. In the event that a node detects an elevated temperature, beyond a user-defined threshold, an alert to nearby nodes is issued to transition to a fire-tracking mode and report the temperature every 10 seconds. We have also developed a Building Monitor application. Within this application, sensor nodes synchronize hourly with the base station to verify that the node is still functioning as well as obtain which operation mode the node should be in. In the low-power mode, nodes do not need to detect movement and return to a sleep state until the next synchronization. During the monitor mode, if the standard deviation of the last four samples is greater than a user-defined threshold, the sensor node will transmit a message to the base station indicating movement with the corresponding time. In contrast to the forest fire monitoring application that is *periodic* in nature, the building monitor is a more *reactive* system in that it reports movement when detected.

We implemented all applications without profiling and with each of the four profiling methods to determine the network traffic, energy, and code size overheads. Results are presented in Table 1.1.

TABLE 1.1

Network Traffic (%/bytes), Energy Consumption (mAh), and Code Size (%/kb) Overheads of Various Profiling Strategies for the Forest Fire Detection and Building Monitor Applications

	Profiling Method PM1			Profiling Method PM2		
Application	Traffic	Energy	Code	Traffic	Energy	Code
Forest fire	8.7%/35	0.01	1.9%/0.8	14.8%/98	0.06	2.2%/0.9
Building monitor	11.6%/18	0.02	3.5%/1.4	32.2%/30	0.02	2.5%/1.0

	Profiling Method PM3			Profiling Method PM4		
Application	Traffic	Energy	Code	Traffic	Energy	Code
Forest fire	7.9%/54	0.02	2.4%/1.0	8.4%/50	<0.01	1.7%/0.7
Building monitor	17.8%/16	0.01	3.0%/1.2	16.5%/17	<0.01	2.2%/0.9

For the forest fire detection and propagation tracking application, PM3 yields the lowest network traffic overhead. Although the profile information is sent as individual packets within the PM3 strategy, by eliminating the need to transmit profile request packets from the base station, the overhead is reduced compared to both PM1 and PM2. However, for the building monitor application, the PM1 strategy incurs the lowest overhead. For this application, the profiling data is primarily piggybacked within the time-synchronization packets. Due to the low overall number of packets transmitted within the network for this application, piggybacking significantly reduces the additional traffic that would be required for the other methodologies. This implies that profiler-module-directed profiling is well suited to reactive systems due to the unpredictable nature of these applications. In addition, piggybacking profile data to existing packets is only preferable when periodic profiling is employed, as transmitting both profile request packets and separate profile packets leads to significant 14.8% and 32.2% overheads for the two respective applications.

Across all profiling methods, energy and code size overhead remain reasonable, with a maximum overhead of only 0.06 milliamp-hours and 1.4 kilobytes (or 3.5%).

1.6 DYNAMIC OPTIMIZATIONS

1.6.1 DYNAMIC OPTIMIZATION METHODOLOGY

Figure 1.3 depicts our dynamic optimization methodology for WSNs, where all operations within the larger shaded circle correspond to the DPOP *optimizer* in Figure 1.1, with the exception of the *dynamic profiler*, which corresponds to the DPOP *profiler* in Figure 1.1. WSN designers evaluate application requirements and capture these requirements as high-level application metrics (e.g., lifetime, throughput, reliability) and associated weight factors. The weight factors characterize the relative importance of the application metrics. (For example, since some WSN applications may not be power-centric and throughput may be more important than the lifetime, assigning a higher weight factor for throughput than lifetime can capture this relationship.) The sensor nodes use application metrics and weight factors to determine an appropriate operating state (tunable parameter settings) using an application metric estimation model. This model determines high-level application metric values corresponding to the tunable parameter settings.

Figure 1.3 shows the per-node dynamic optimization process (encompassed by the dashed circle), which is orchestrated by the *dynamic optimization controller*. The process consists of two operating modes: the *one-shot mode*, wherein the sensor-node operating state is directly determined, and the *improvement mode*, wherein the operating state is iteratively improved using an online optimization algorithm. The dynamic optimization process consists of three steps. In the first step, the dynamic optimization controller intelligently determines the initial parameter-value settings (operating state) and exploration order (ascending or descending), which is critical in reducing the number of states explored by the improvement mode. In the one-shot operation mode, the dynamic optimization process is complete and the sensor node moves directly to the operating state specified by the initial parameter-value settings.

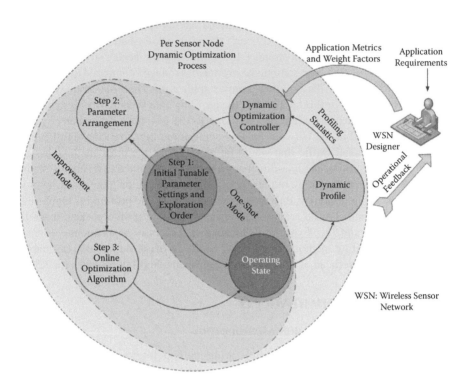

FIGURE 1.3 Our dynamic optimization methodology for wireless sensor networks. All operations within the larger shaded circle correspond to the DPOP *optimizer* in Figure 1.1, with the exception of the *dynamic profiler*, which corresponds with the DPOP *profiler* in Figure 1.1.

In the improvement mode, the second step determines the parameter arrangement based on weight factors for the application metrics (e.g., explore processor voltage then frequency then sensing frequency). The third step invokes an *online optimization algorithm* for parameter exploration to iteratively improve the operating state to more closely meet application requirements. The online optimization algorithm leverages the intelligent initial parameter-value settings, exploration order, and parameter arrangement. A *dynamic profiler* records profiling statistics (e.g., processor voltage, wireless channel condition, radio transmission power) given the current operating state and environmental stimuli, and then passes these profiling statistics to the dynamic optimization controller.

The dynamic optimization controller processes the profiling statistics to determine whether the current operating state meets the application requirements. If the application requirements are not met, the dynamic optimization controller reinvokes the dynamic optimization process to determine a new operating state. This feedback process continues to ensure that the application requirements are best met under changing environmental stimuli.

1.6.2 DYNAMIC OPTIMIZATION FORMULATION

In the following section, we formulate the state space and the objective function utilized within our dynamic optimization framework.

1.6.2.1 State Space

The state space S for our dynamic optimization methodology given N tunable parameters is defined as:

$$S = P_1 \times P_2 \times \ldots \times P_N, \tag{1.1}$$

where P_i denotes the state space for tunable parameter i, $\forall\, i \in \{1,2,\ldots,N\}$ and $\times$ denotes the Cartesian product. Each tunable parameter P_i consists of n values:

$$P_i = \{p_{i_1}, p_{i_2}, p_{i_3}, \ldots,\ p_{i_n}\} : |P_i| = n \tag{1.2}$$

where $|P_i|$ denotes the tunable parameter P_i state-space cardinality (the number of tunable values in P_i). S is a set of n-tuples formed by taking one value from each tunable parameter. A single n-tuple $s \in S$ is given as

$$s = (p_{1_y}, p_{2_y}, \ldots, p_{N_y}) : p_{i_y} \in P_i, \quad \forall\, i \in \{1,2,\ldots,N\}, \quad y \in \{1,2,\ldots,n\} \tag{1.3}$$

Each n-tuple represents a sensor-node operating state. We point out that some n-tuples in S may not be feasible (such as invalid combinations of processor voltage and frequency) and can be regarded as *do not care* tuples.

1.6.2.2 Optimization Objective Function

The dynamic optimization problem can be formulated as

$$
\begin{aligned}
\max\ f(s) &= \sum_{k=1}^{m} \omega_k f_k(s) \\
s.t.\ \ s &\in S \\
\omega_k &\geq 0,\ k = 1,2,\ldots,\ m \\
\omega_k &\leq 1,\ \ k = 1,2,\ldots,\ m\,, \\
\sum_{k=1}^{m} \omega_k &= 1
\end{aligned}
\tag{1.4}
$$

where $f(s)$ denotes the objective function that captures application metrics and weight factors. In Equation (1.4), $f_k(s)$ and ω_k denote the objective function and weight factor for the kth application metric, respectively, given that there are m application metrics. Each state $s \in S$ has an associated objective-function value, and the optimization goal is to determine a state that gives maximum (optimal) objective-function value $f^{\mathrm{opt}}(s)$, which indicates the best possible adherence to specified application requirements given the design space S. The solution quality for any $s \in S$ can be determined by normalizing the objective-function value corresponding to state s with respect to

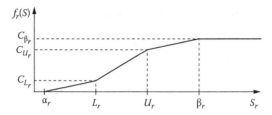

FIGURE 1.4 Reliability objective function $f_r(s)$.

$f^{opt}(s)$. The normalized objective-function value corresponding to a state can vary from 0 to 1, where 1 indicates the optimal solution.

For our dynamic optimization methodology, we consider three application metrics ($m = 3$)—lifetime, throughput, and reliability—whose objective functions are denoted by $f_l(s)$, $f_t(s)$, and $f_r(s)$, respectively. We define $f_r(s)$ (Figure 1.4) using the piecewise linear function:

$$f_r(s) = \begin{cases} 1, & s_r \geq \beta_r \\[2ex] C_{U_r} + \dfrac{(C_{\beta_r} - C_{U_r})(s_r - U_r)}{(\beta_r - U_r)}, & U_r \leq s_r < \beta_r \\[2ex] C_{L_r} + \dfrac{(C_{U_r} - C_{L_r})(s_r - L_t)}{(U_r - L_r)}, & L_r \leq s_r < U_r \\[2ex] C_{L_r} \cdot \dfrac{(s_r - \alpha_r)}{(L_r - \alpha_r)}, & \alpha_r \leq s_r < L_r \\[2ex] 0, & s_t < \alpha_t \end{cases} \tag{1.5}$$

where s_r denotes the reliability offered by state s, the constant parameters L_r and U_r denote the *desired* minimum and maximum reliability, and the constant parameters α_r and β_r denote the *acceptable* minimum and maximum reliability. The piecewise linear objective function provides WSN designers with a flexible application requirement specification, as it allows both desirable and acceptable ranges. The objective-function reward gradient (slope) would be greater in the desired range than the acceptable range; however, there would be no reward for operating outside the acceptable range. The constant parameters C_{L_r}, C_{U_r}, and C_{β_r} in Equation (1.5) denote the $f_r(s)$ value at L_r, U_r, and β_r, respectively.

The $f_l(s)$ and $f_t(s)$ functions can be defined using increasing piecewise linear functions similar to Equation (1.5), as higher values of lifetime and throughput (like reliability) are typically desirable and correspond to higher objective-function values. Although we define our objective functions using piecewise linear objective functions, our dynamic optimization methodology works well for any other characterization of objective functions (e.g., linear, nonlinear).

1.6.3 Dynamic Optimization Algorithms

In this section, we describe the three steps, associated algorithms, and operating modes for our dynamic optimization methodology (Section 1.6.1). Step one determines initial tunable parameter values and exploration order (ascending or descending). In *one-shot mode* (Figure 1.3), these initial tunable parameter-value settings result in a high-quality operating state in a one-shot solution (no additional design-space exploration) for applications with tight constraints (e.g., limited exploration time due to a rapidly changing environment). For applications with more flexible constraints, the *improvement mode* encompasses steps two and three and iteratively improves the one-shot solution. Step two determines the tunable parameter-exploration arrangement based on the weight factors for the application metrics (i.e., some parameters are more critical for an application than others and should be explored first). Step three leverages the outcomes of steps one and two and explores the design space using an online optimization algorithm. For this optimization algorithm, we propose a lightweight greedy algorithm for design-space exploration. We note, however, that step three can be generalized to any online algorithm.

1.6.3.1 Initial Parameter-Value Settings and Exploration Order

Algorithm 1.1 describes our technique to determine initial tunable parameter-value settings and exploration order (first step of our dynamic optimization methodology). The algorithm takes as input the objective function $f(s)$, the number of tunable parameters N, the number of values for each tunable parameter n, the number of application

```
1   for k ← 1 to m do
2   │  for P_i ← P_1 to P_N do
3   │  │   f^k_{k_{i1}} ← k-metric objective function value when parameter
    │  │   setting is {P_i = p_{i_1}, P_j = P_{j0}, ∀ i ≠ j};
4   │  │   f^k_{p_{i_n}} ← k-metric objective function value when parameter
    │  │   setting is {P_i = p_{i_n}, P_j = P_{j0}, ∀ i ≠ j};
5   │  │   δf^k_{P_i} ← f^k_{p_{i_n}} − f^k_{p_{i1}};
6   │  │   if δf^k_{P_i} > 0 then
7   │  │   │   explore P_i in descending order;
8   │  │   │   p^k_k[i] ← descending;
9   │  │   │   p^k_0[i] ← p^k_{i_n};
10  │  │   else
11  │  │   │   explore P_i in ascending order;
12  │  │   │   P^k_d[i] ← ascending;
13  │  │   │   P^k_0[i] ← p^k_{i_1};
14  │  │   end
15  │  end
16  end
        return P^k_d, P^k_0, ∀κ ∈ {1, …, m}
```

ALGORITHM 1.1 Initial tunable parameter-value settings and exploration-order algorithm.

metrics m, and P, where P represents a vector containing the tunable parameters $P = \{P_1, P_2, \ldots, P_N\}$. For each application metric k, the algorithm calculates vectors P_0^k and P_d^k (where d denotes the exploration direction [ascending or descending]), which store the initial value settings and exploration order, respectively, for the tunable parameters. The algorithm determines the kth application metric objective-function values $f_{P_{i_1}}^k$ and $f_{P_{i_n}}^k$, where the parameter being explored, P_i, is assigned its first P_{i_1} and last P_{i_n} tunable values, respectively, and the rest of the tunable parameters, P_j, $\forall \, j \neq i$, are assigned initial values (lines 3–4). The function $\delta f_{P_i}^k$ stores the difference between $f_{P_{i_n}}^k$ and $f_{P_{i_1}}^k$. For $\delta f_{P_i}^k > 0$, P_{i_n} results in a greater objective-function value as compared to p_i for parameter P_i (i.e., the objective-function value decreases as the parameter value decreases). Therefore, to reduce the number of states explored while considering that the greedy algorithm (Section 1.6.3.3) will stop exploring a tunable parameter if a tunable parameter's value yields a comparatively lower objective-function value, P_i's exploration order must be descending (lines 6–8). The algorithm assigns P_{i_n} as the initial value of P_i for the kth application metric (line 9). If $\delta f_{P_i}^k < 0$, the algorithm assigns the exploration order as ascending for P_i and P_{i_1} as the initial value setting of P_i (lines 11–13). This $\delta f_{P_i}^k$ calculation procedure is repeated for all m application metrics and all N tunable parameters (lines 1–16).

1.6.3.2 Parameter Arrangement

Depending on the weight factors for the application metrics, some parameters are more critical to meeting application requirements than other parameters. For example, sensing frequency is a critical parameter for applications with a high responsiveness weight factor and, therefore, sensing frequency should be explored first. In this subsection, we devise a technique for parameter arrangement such that parameters are explored in an order characterized by the parameter's impact on application metrics based on relative weight factors. Our parameter-arrangement technique is based on calculations performed in Algorithm 1.1. We define

$$\nabla f_P = \left\{ \nabla f_P^1, \nabla f_P^2, \ldots, \nabla f_P^m \right\}, \tag{1.6}$$

where ∇f_p is a vector containing $\nabla f_P^k, \forall \, k \in \{1, 2, \ldots, m\}$ arranged in descending order by their respective values and is given as

$$\nabla f_P^k = \left\{ \delta f_{P_1}^k, \delta f_{P_2}^k, \ldots, \delta f_{P_N}^k \right\} \; : \; \left| \delta f_{P_i}^k \right| \geq \left| \delta f_{P_{i+1}}^k \right|, \forall \, i \in \{1, 2, \ldots, N-1\}. \tag{1.7}$$

The tunable parameter-arrangement vector P^k corresponding to ∇f_P^k (one-to-one correspondence) is given by

$$P^k = \left\{ P_1^k, P_2^k, \ldots, P_N^k \right\}, \forall \, k \in \{1, 2, \ldots, m\}. \tag{1.8}$$

An intelligent parameter arrangement $\hat{P}$ must consider all application metrics' weight factors, with higher importance given to higher weight factors, i.e.,

$$\hat{P} = \left\{ P_1^1, \ldots, P_{l_1}^1, P_1^2, \ldots, P_{l_2}^2, P_1^3, \ldots, P_{l_3}^3, \ldots, P_1^m, \ldots, P_{l_m}^m \right\}, \tag{1.9}$$

where l_k denotes the number of tunable parameters taken from $P^k, \forall\ k \in \{1,2,...,m\}$ such that $\sum_{k=1}^{m} l_k = N$. Our technique allows taking more tunable parameters from parameter-arrangement vectors corresponding to application metrics with a higher weight factor (i.e., $l_k \geq l_{k+1}, \forall\ k \in \{1,2,...,m-1\}$). In Equation (1.9), l_1 tunable parameters are taken from vector P^1, then l_2 from vector P^2, and so on to l_m from vector P^m such that $\{P_1^k,...,P_{l_k}^k\} \cap \{P_1^{k-1},...,P_{l_{k-1}}^{k-1}\} = \emptyset, \forall\ k \in \{2,3,...,m\}$. In other words, we select those tunable parameters from parameter-arrangement vectors corresponding to lower weight factors that are not already selected from parameter-arrangement vectors corresponding to higher weight factors (i.e., $\hat{P}$ comprises disjoint or non-overlapping tunable parameters corresponding to each application metric).

In the situation where weight factor ω_1 is much greater than all other weight factors, an intelligent parameter arrangement $\tilde{P}$ would correspond to the parameter arrangement for the application metric with weight factor ω_1, i.e.,

$$\tilde{P} = P^1 = \left\{ P_1^1, P_2^1, ..., P_N^1 \right\} \Leftrightarrow \omega_1 \gg \omega_q, \quad \forall\ q \in \{2,3,...,m\} \tag{1.10}$$

The initial parameter-value vector $\hat{P}_0$ and exploration-order (ascending or descending) vector $\hat{P}_d$ corresponding to $\hat{P}$ can be determined using Equation (1.9), P_d^k, and $P_0^k, \forall\ k \in \{1,...,m\}$ (Algorithm 1.1) by looking at the tunable parameter from P and finding the tunable parameter's initial value from P_0^k and exploration order from P_d^k.

1.6.3.3   Online Optimization Algorithm

The third step of our dynamic optimization process uses a greedy lightweight online optimization algorithm for tunable parameter exploration in an effort to determine a better operating state than the one obtained from step one (Section 1.6.3.1). Algorithm 1.2 depicts our online optimization algorithm, which leverages the initial parameter-value settings (Section 1.6.3.1), parameter-value exploration order (Section 1.6.3.1),

---

Input: $f(s)$, N, n, P, $\hat{P}_0$, $\hat{P}_d$
Output: Sensor node state that maximizes $f(s)$ and the
        corresponding $f(s)$ value

1   $\kappa \leftarrow$ initial tunable parameter value settings from $\hat{P}_0$;
2   $f_{best} \leftarrow$ solution from initial parameter settings κ;
3   **for** $\hat{P}_i \leftarrow \hat{P}_1$ **to** P_N **do**
4 | explore $\hat{P}_i$ in ascending/descending order suggested by $\hat{P}_d$;
5 | **foreach** $\hat{P}_i = \{\hat{P}_{i_1}, \hat{P}_{i_2},...\hat{P}_{i_n}\}$ **do**
6 | | $f_{temp} \leftarrow$ current state ζ solution;
7 | | **if** $f_{temp} > f_{best}$ **then**
8 | | | $f_{best} \leftarrow f_{temp}$;
9 | | | $\xi \leftarrow \zeta$
10 | | **else**
11 | | | break;
12 | | **end**
13 | **end**
14 **end**
    return ξ, f_{best}

ALGORITHM 1.2   Online optimization algorithm for tunable parameter exploration.

and parameter arrangement (Section 1.6.3.2). The algorithm takes as input the objective function $f(s)$, the number of tunable parameters N, the number of values for each tunable parameter n, the tunable parameter's vector P, the tunable parameter's initial-value vector $\hat{P}$, and the tunable parameter's exploration-order (ascending or descending) vector $\hat{P}_d$. The algorithm initializes state κ from $\hat{P}_0$ (line 1) and f_{best} with κ's objective-function value (line 2). The algorithm explores each parameter in $\hat{P}_i$ from Equation (1.9) in ascending or descending order as given by $\hat{P}_d$ (lines 3–4). For each tunable parameter $\hat{P}_i$ (line 5), the algorithm assigns f_{temp}, the objective-function value from the current state ζ (line 6). If $f_{temp} > f_{best}$ (increase in objective-function value), f_{temp} is assigned to f_{best}, and the state ζ is assigned to state ξ (lines 7–9). If $f_{temp} \le f_{best}$, the algorithm stops exploring the current parameter $\hat{P}_i$ and starts exploring the next tunable parameter (lines 10–12). The algorithm returns the best found objective-function value f_{best} and the state ξ corresponding to f_{best}.

1.6.3.4   Computational Complexity

The computational complexity (running time and storage) for our dynamic optimization methodology is $O(Nm \log N + Nn)$, which comprises: the intelligent initial parameter-value settings and exploration ordering (Algorithm 1.1) $O(Nm)$; parameter arrangement $O(Nm \log N)$ (sorting ∇f_p^k from Equation (1.7) contributes the $N \log N$ factor) (Section 1.6.3.2); and the online optimization algorithm for parameter exploration (Algorithm 1.2) $O(Nn)$. Assuming that the number of tunable parameters N is larger than the number of parameters' tunable values n, the computational complexity of our methodology can be given as $O(Nm \log N)$. This complexity reveals that our proposed methodology is lightweight and is thus feasible for implementation on sensor nodes with tight resource constraints.

1.6.4   EXPERIMENTAL RESULTS

1.6.4.1   Experimental Setup

Our experimental setup is based on the Crossbow IRIS mote platform [62] with a battery capacity of 2,000 mA-h using two AA alkaline batteries. The IRIS mote platform integrates an Atmel ATmega1281 microcontroller [63], an MTS400 sensor board [64] with Sensirion SHT1x temperature and humidity sensors [65], and an Atmel AT-86RF230 low-power 2.4-GHz transceiver [66].

We analyze six tunable parameters: processor voltage V_p, processor frequency F_p, sensing frequency F_s, packet size P_s, packet transmission interval P_{ti}, and transceiver transmission power P_{tx}. In order to explore the fidelity of our methodology across small and large design spaces, we consider two design-space cardinalities (number of states in the design space) $|S| = 729$ and $|S| = 31,104$. The tunable parameters for $|S| = 729$ are

$V_p = \{2.7, 3.3, 4\}$ (volts)
$F_p = \{4, 6, 8\}$ (MHz) [63]
$F_s = \{1, 2, 3\}$ (samples per second) [65]

$P_s = \{41, 56, 64\}$ (bytes)
$P_{ti} = \{60, 300, 600\}$ (seconds)
$P_{tx} = \{-17, -3, 1\}$ (dBm) [66]

The tunable parameters for $|S| = 31,104$ are

$V_p = \{1.8, 2.7, 3.3, 4, 4.5, 5\}$ (volts)
$F_p = \{2, 4, 6, 8, 12, 16\}$ (MHz) [63]
$F_s = \{0.2, 0.5, 1, 2, 3, 4\}$ (samples per second) [65]
$P_s = \{32, 41, 56, 64, 100, 127\}$ (bytes)
$P_{ti} = \{10, 30, 60, 300, 600, 1200\}$ (seconds)
$P_{tx} = \{-17, -3, 1, 3\}$ (dBm) [66]

All state-space tuples are feasible for $|S| = 729$, whereas $|S| = 31,104$ contains 7,779 infeasible state-space tuples (i.e., not all V_p and F_p pairs are feasible).

In order to evaluate the robustness of our methodology across different applications with varying weight factors for application metrics, we model three sample application domains (a security/defense system, a health-care application, and an ambient conditions monitoring application) and assign application-specific values for the desirable minimum L, desirable maximum U, acceptable minimum α, and acceptable maximum β objective-function parameter values for application metrics (Section 1.6.2.2). Our selected objective-function parameter values and application-metrics weight factors represent typical application requirements [1]. Although we analyzed our methodology for the IRIS motes platform, three application domains, and two design spaces, our algorithms are equally applicable to any platform, application domain, and design space.

1.6.4.2   Results

For comparison purposes, we implemented a simulated annealing (SA)–based algorithm, our greedy online optimization algorithm (GD) (which leverages intelligent initial parameter-value selection, exploration ordering, and parameter arrangement), and several other greedy online algorithm variations (Table 1.2) in C/C++. First, we evaluated our methodology and algorithms with a desktop implementation to reduce the analysis time and study the feasibility of our algorithms for a dynamic environment. We compared our results with SA to provide relative comparisons of greedy algorithms with another heuristic algorithm. We compared GD results with different greedy-algorithm variations (Table 1.2) to provide an insight into how initial parameter-value settings, exploration ordering, and parameter arrangement affect the final operating state quality. We normalized the objective-function value (corresponding to the operating state) attained by the algorithms with respect to the optimal solution obtained using an exhaustive search. We analyzed the relative complexity of the algorithms by measuring the execution time and data memory requirements. Note that for brevity, our presented results are a subset of the greedy algorithms listed in Table 1.2, the application domains, and the design spaces. However, we evaluated all the greedy algorithms and application domains, and the subset presented in this section is representative of all greedy algorithms and application-domain trends and characteristics.

TABLE 1.2

Greedy Algorithms with Different Parameter Arrangements and Exploration Orders

Notation	Description
GD	Our greedy algorithm with parameter-exploration order $\hat{P}_d$ and arrangement $\hat{P}$
GD^{ascA}	Explores parameter values in ascending order with arrangement $A = \{V_p, F_p, F_s, P_s, P_{ti}, P_{tx}\}$
GD^{ascB}	Explores parameter values in ascending order with arrangement $B = \{P_{tx}, P_{ti}, P_s, F_s, F_p, V_p\}$
GD^{ascC}	Explores parameter values in ascending order with arrangement $C = \{F_s, P_{ti}, P_{tx}, V_p, F_p, P_s\}$
GD^{desD}	Explores parameter values in descending order with arrangement $D = \{V_p, F_p, F_s, P_s, P_{ti}, P_{tx}\}$
GD^{desE}	Explores parameter values in descending order with arrangement $E = \{P_{tx}, P_{ti}, P_s, F_s, F_p, V_p\}$
GD^{desF}	Explores parameter values in descending order with arrangement $F = \{P_s, F_p, V_p, P_{tx}, P_{ti}, F_s\}$

Figure 1.5 shows the objective-function value normalized to the optimal solution versus number of states explored for a security/defense system for |S| = 729. GD^{ascA}, GD^{desD}, and GD converge to a steady-state solution after exploring 11, 10, and 8 states, respectively. These convergence results show that GD converged to the optimal solution slightly faster than other greedy algorithms, exploring only 1.1% of the design space. The order in which greedy algorithms explore tunable parameters (e.g., V_p, then F_p, and so on) affects the number of iterations for convergence similarly to how the order of variables in a binary decision diagram (BDD) impacts the size of the tree and processing speed [67]. GD^{ascA} converges to a better solution than GD^{desD}, showing that ascending parameter-values exploration is better for a security/defense

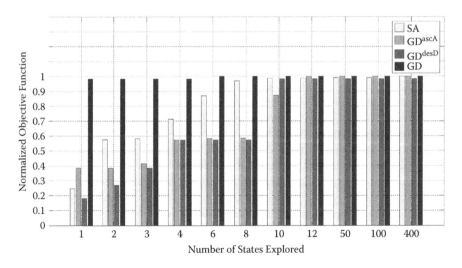

FIGURE 1.5 Objective-function value normalized to the optimal solution for a varying number of states explored for SA and the greedy algorithms for a security/defense system, where $\omega_l = 0.25$, $\omega_t = 0.35$, $\omega_r = 0.4$, |S| = 729.

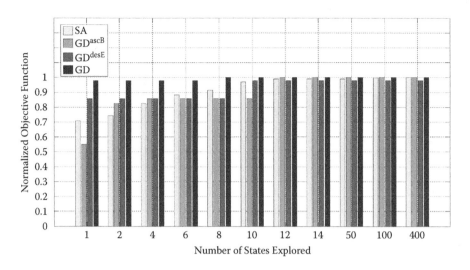

FIGURE 1.6 Objective-function value normalized to the optimal solution for a varying number of states explored for SA and greedy algorithms for a health-care application, where $\omega_l = 0.25$, $\omega_t = 0.35$, $\omega_r = 0.4$, $|S| = 729$.

system with the given weight factors. In addition, the SA algorithm outperforms all greedy algorithms and converges to the optimal solution for $|S| = 729$ after exploring 400 states (54.9% of the design space). Figure 1.5 also verifies the ability of our methodology to determine a good quality (near-optimal), one-shot solution, as GD achieves only a 1.83% improvement over the initial state after exploring eight states.

Figure 1.6 shows the objective-function value normalized to the optimal solution versus number of states explored for a health-care application for $|S| = 729$. Figure 1.6 shows similar trends as seen in Figure 1.5 for convergence rates on all algorithms, GD's one-shot solution quality, and GD^{ascB}'s superior solution as compared to GD^{desE}.

Figure 1.7 shows the objective-function value normalized to the optimal solution versus number of states explored for an ambient-conditions-monitoring application for $|S| = 31,104$. GD converges to the optimal solution after exploring 13 states (0.04% of the design space), with a 16.69% improvement over the one-shot solution. GD^{ascC} and GD^{desF} converge to the solution after exploring 9 and 7 states, respectively. These convergence and percentage improvement results show that GD may explore more states than other greedy algorithms if state exploration provides a noticeable improvement over the one-shot solution. The figure also shows that SA converges to a near-optimal solution after exploring 400 states (1.29% of the design space). These convergence results show that even though the design-space cardinality increases by 43×, both heuristic algorithms (greedy and SA) still explore only a small percentage of the design space and result in high-quality solutions.

In order to prove the effectiveness of our intelligent initial parameter-value selection technique (Section 1.6.3.1), we calculated the percentage improvements in the normalized objective-function value obtained by the intelligent initial parameter-value settings over other arbitrary initial value settings used in other greedy

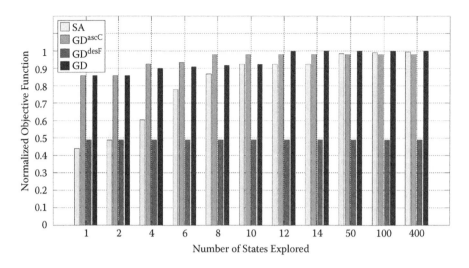

FIGURE 1.7 Objective-function value normalized to the optimal solution for a varying number of states explored for SA and greedy algorithms for an ambient-conditions-monitoring application, where $\omega_l = 0.6$, $\omega_t = 0.25$, $\omega_r = 0.15$, $|S| = 31{,}104$.

algorithms for $|S| = 729$ (Table 1.3) and $|S| = 31{,}104$ (Table 1.4). Results reveal that our one-shot operating state using only intelligent initial parameter settings is within 5.92% of the optimal averaged over several different application domains and design spaces. Table 1.3 and Table 1.4 verify that our intelligent initial parameter settings technique (which gives a near-optimal solution) provides substantial percentage improvements over other arbitrary initial settings for different application domains and design-space cardinalities. We point out that some arbitrary initial parameter settings may attain a slightly higher normalized objective-function value for a particular application and design-space cardinality (e.g., initial parameter settings for GD^{desE} for the ambient-conditions-monitoring application when $|S| = 31{,}104$), but in general, the arbitrary selection would not scale to other applications and design-space cardinalities.

TABLE 1.3
Improvement Attained by GD (%) after One-State Exploration when S = 729

Application Domain	GD^{ascA}	GD^{ascB}	GD^{ascC}	GD^{desD}	GD^{desE}	GD^{desF}
Security/defense system	155.06	155.06	155.06	439.56	22.9	9.2
Health care	78	78	78	211.78	14.1	6.64
Ambient conditions monitoring	51.81	51.81	51.81	142.89	39.85	6.09

TABLE 1.4

Improvement Attained by GD (%) after One-State Exploration when S = 31,104

Application Domain	GD^{ascA}	GD^{ascB}	GD^{ascC}	GD^{desD}	GD^{desE}	GD^{desF}
Security/ defense system	147.97	147.97	147.97	435.09	13.66	0.33
Health care	73.42	73.42	73.42	218.77	11.47	0.3
Ambient conditions monitoring	0	0	0	146.97	−8.24	75.61

We performed data memory analysis for each step of our dynamic optimization methodology (Section 1.6.1). Step one (one-shot solution) requires only 150, 188, 248, and 416 bytes, whereas step two requires 94, 140, 200, and 494 bytes for (number of tunable parameters N, number of application metrics m) equal to (3, 2), (3, 3), (6, 3), and (6, 6), respectively. For step three, we compared data memory requirements for GD with SA for different design-space cardinalities. We observed that GD requires 458, 528, 574, 870, and 886 bytes, whereas SA requires 514, 582, 624, 920, and 936 bytes of storage for design-space cardinalities of 8, 81, 729, 31,104, and 46,656, respectively. The data memory analysis shows that SA has comparatively larger memory requirements than the greedy algorithm. Our analysis reveals that the data memory requirements for all three steps of our dynamic optimization methodology increase linearly as the number of tunable parameters, tunable values, and application metrics (and thus the design space) increases. Furthermore, the data memory analysis verifies that although our dynamic optimization methodology (all three steps) has low data memory requirements, the one-shot solution (from step one) requires 361% less memory on average.

We measured the execution time for all three steps of our dynamic optimization methodology averaged over 10,000 runs (to smooth any discrepancies in execution time due to operating system overheads) on an Intel Xeon CPU running at 2.66 GHz [68] using the Linux/Unix *time* command [69]. We scaled these execution times to the Atmel ATmega1281 microcontroller [63] running at 8 MHz. Even though scaling does not provide 100% accuracy for the microcontroller runtime because of different instruction set architectures, scaling provides reasonable runtime estimates and enables relative comparisons.

Results showed that step one and step two required 1.66 ms and 0.332 ms, respectively, both for |S| = 729 and |S| = 31,104. For step three, we compared GD with SA. GD explored 10 states and required 0.887 ms and 1.33 ms on average to converge to the solution for |S| = 729 and |S| = 31,104, respectively. SA took 2.76 ms and 2.88 ms to explore the first 10 states (to provide a fair comparison with GD) for |S| = 729 and |S| = 31,104, respectively. The other greedy algorithms required comparatively more time than GD because they required more state explorations to converge than GD.

However, all greedy algorithms required less execution time than SA. To verify that our dynamic optimization methodology is lightweight, we compared the execution time results for our dynamic optimization methodology (including all three steps) with the exhaustive search. The exhaustive search required 29.526 ms and 2.765 s for $|S| = 729$ and $|S| = 31,104$, respectively. Compared with the exhaustive search, our dynamic optimization methodology required 10× and 832× less execution time for $|S| = 729$ and $|S| = 31,104$, respectively.

The execution time analysis reveals that our dynamic optimization methodology (including all three steps) requires execution time on the order of milliseconds, and the one-shot solution requires 138% less execution time on average as compared to all three steps of the dynamic optimization methodology. Execution time savings attained by the one-shot solution as compared to the three steps of our dynamic optimization methodology are 73% and 186% for GD and SA, respectively, when $|S| = 729$, and are 100% and 138% for GD and SA, respectively, when $|S| = 31,104$. These results indicate that the design-space cardinality affects the execution time linearly and that our dynamic optimization methodology's advantage increases as design-space cardinality increases.

1.7 CONCLUSIONS

Dynamic profiling of sensor-based platforms enables an accurate view of an application's execution behavior, but it must balance the network traffic, energy, and code size overheads. A variety of dynamic methods have been considered to analyze the corresponding overheads. While the increases in energy and code size are reasonable, network traffic overheads range from 7.9% to 32.2%. Furthermore, choosing an appropriate profiling mechanism is dependent on the application behavior itself, and a single profiling method is unlikely to provide good results across different classes of applications. A classification methodology or tool is needed to select and adjust the profiling method for a given application based on that application's specific characteristics.

To leverage the dynamic profiling information, we proposed a dynamic optimization methodology that provides a high-quality one-shot solution using intelligent initial tunable parameter-value settings for highly constrained applications. Additionally, we proposed an online greedy optimization algorithm that leveraged intelligent design-space exploration techniques to iteratively improve on the one-shot solution for less-constrained applications.

Compared with simulating annealing and different greedy-algorithm variations, results showed that the one-shot technique yielded improvements as high as 440% over other arbitrary initial-parameter settings. Results indicated that our greedy algorithm converged to the optimal (or near-optimal) solution after exploring only 1.1% and 0.04% of the design space, whereas SA explored 54.9% and 1.29% of design spaces with 729 and 31,104 different operating states, respectively. Analysis of data memory and execution time confirmed that our one-shot solution and our entire dynamic optimization methodology are lightweight and thus feasible for use in sensor nodes with limited resourcAcknowledgments

This work was supported by the National Science Foundation (NSF) (CNS-0834080 and CNS-0834102) and Natural Sciences and Engineering Research

Council of Canada (NSERC). Any opinions, findings, and conclusions or recommendations expressed in this material are those of the author(s) and do not necessarily reflect the views of the NSF and NSERC.

REFERENCES

1. Abdelzaher, T., B. Blum, Q. Cao, D. Evans, S. George, J. Stankovic, T. He, et al. 2004. EnviroTrack: Towards an environmental computing paradigm for distributed sensor networks. In *IEEE International Conference on Distributed Computing Systems*, 582–89. Piscataway, NJ: IEEE Press.
2. Akyildiz, I., W. Su, Y. Sankarasubramaniam, and E. Cayirci. 2002. Wireless sensor networks: A survey. *Elsevier Computer Networks* 38 (4): 393–422.
3. Barr, R., J. C. Bicket, D. S. Dantas, B. Du, T. W. Danny Kim, B. Zhou, and E. Gun Sirer. 2002. On the need for system-level support for ad hoc and sensor networks. *Operating System Review* 36 (2): 1–5.
4. Bischoff, U., and G. Kortuem. 2007. A state-based programming model and system for wireless sensor networks. In *International Conference on Pervasive Computing and Communications Workshops*, 261–66. Washington, DC: IEEE Computer Society.
5. Bonnet, P., J. Gehrke, and P Seshardi. 2001. Towards sensor database systems. In *Second International Conference on Mobile Data Management (MDM)*, 3–14. London: Springer-Verlag.
6. Boulis, A., C. Han, and M. B. Srivastava. 2003. Design and implementation of a framework for efficient and programmable sensor networks. In *Proceedings of First International Conference on Mobile Systems, Applications, and Services (MobiSys)*, 187–200. Berkeley, CA: USENIX Association.
7. Boulis, A., and M. B. Srivastava. 2002. A framework for efficient and programmable sensor networks. Paper presented at IEEE Open Architectures and Networking Programming Conference (OpenArch). New York, NY.
8. Ghercioiu, M. 2005. A graphical programming approach to wireless sensor network nodes. In *Sensors for Industry Conference Proceedings*, 118–21. Piscataway, NJ: IEEE Press.
9. Girod, L., J. Elson, A. Cerpa, T. Stathopoulos, N. Ramanathan, and D. Estrin. 2004. EmStar: A software environment for developing and deploying wireless sensor networks. In *USENIX Technical Conference Proceedings*, 283–96. Berkeley, CA: USENIX Association.
10. Hazelwood, K., and M. Smith. 2006. Managing bounded code caches in dynamic binary optimization systems. *ACM Trans. Architecture and Code Optimization* 3 (3): 263–94.
11. Horey, J., E. Nelson, and A. B. Maccabe. 2007. Tables: A table-based language environment for sensor networks. Technical report. The University of New Mexico.
12. Kogekar, S., S. Neema, B. Eames, X. Koutsoukos, A. Ledeczi, and M. Maroti. 2004. Constraint-guided dynamic reconfiguration in sensor networks. In *International Symposium on Information Processing in Sensor Networks (IPSN'04)*, 379–87. New York: ACM Press.
13. Levis, P., D. Gay, and D. Culler. 2004. Bridging the gap: Programming sensor networks with application specific virtual machines. Submitted to Operating Systems Design and Implementation Conference (OSDI 04).
14. Levis, P., N. Lee, M. Welsh, and D. Culler. 2003. TOSSIM: Accurate and scalable simulation of entire TinyOS applications. In *Conference on Embedded Networked Sensor Systems (SenSys)*, 126–37. New York: ACM Press.
15. Madden, S., M. Franklin, J. Hellerstein, and W. Hong. 2005. TinyDB: An acquisitional query processing system for sensor metworks. *ACM Trans. Database Systems* 30 (1): 122–73.

16. Mainland, G., L. Kang, S. Lahaie, D. C. Parkes, and M. Welsh. 2004. Using virtual markets to program global behavior in sensor networks. In *Proceedings of the 11th ACM SIGOPS European Workshop*, article 1. New York: ACM Press.

17. Mannion, R., H. Hsieh, S. Cotterell, and F. Vahid. 2005. System synthesis for networks of programmable blocks. In *Conference on Design, Automation and Test in Europe (DATE)*, 888–93. Piscataway, NJ: IEEE Press.

18. Muller, R., G. Alonso, and D. Kossmann. 2007. SwissQM: Next generation data processing in sensor networks. In *Conference on Innovative Data Systems Research (CIDR)*, 1–9. http://www.cidrdb.org/cidr2007/index.html.

19. Newton, R., G. Morrisett, and M. Welsh. 2007. The regiment macroprogramming system. In *International Conference on Information Processing in Sensor Networks (IPSN)*, pp. 489–498. New York: ACM Press.

20. Newton, R., and M. Welsh. 2004. Region streams: Functional macroprogramming for sensor networks. In *First International Workshop on Data Management for Sensor Networks (DMSN)*, 78–87. New York: ACM Press.

21. Srisathapornphat, C., C. Jaikeo, and C. Shen. 2001. Sensor information networking architecture and applications. *IEEE Personal Communications* 8 (4): 52–59.

22. Yao, Y., and J. E. Gehrke. 2002. The cougar approach to in-network query processing in sensor networks. *ACM SIGMOD Record* 31 (3): 9–18.

23. Welsh, M., and G. Mainland. 2004. Programming sensor networks using abstract regions. In *Proceedings of the First USENIX/ACM Symposium on Networked Systems Design and Implementation (NSDI '04)*, 3–3. Berkeley, CA: USENIX Association.

24. Hill, J., and D. Culler. 2002. MICA: A wireless platform for deeply embedded networks. *IEEE Micro.* 22 (6): 12–24.

25. Dutta, P., and D. Culler. 2005. System software techniques for low-power operation in wireless sensor networks. In *Proceedings of International Conference on Computer-Aided Design (ICCAD)*, 925–32. Washington, DC: IEEE Computer Society.

26. Kadayif, I., and M. Kandemir. 2004. Tuning in-sensor data filtering to reduce energy consumption in wireless sensor networks. In *Proceedings of Design, Automation, and Test in Europe (DATE) Conference*, 1530–91. Piscataway, NJ: IEEE Press.

27. Polley, J., D. Blazakis, J. McGee, D. Rusk, J. S. Baras, and M. Karir. 2004. ATEMU: A fine-grained sensor network simulator. In *Conference on Sensor and Ad Hoc Communications and Networks (SECON)*, 477–82. New York: IEEE Communications Society.

28. Perrone, F., and D. Nicol. 2002. A scalable simulator for TinyOS applications. In *Proceedings of the Winter Simulation Conference*, 679–87. Piscataway, NJ: IEEE Press.

29. Yu, Y., D. Ganseen, L. Girod, D. Estrin, and R. Govindan. Synthetic data generation to support irregular sampling in sensor networks. *GeoSensor Networks* 1 (4): 211–34.

30. Park, C., and P. Chou. 2006. EmPro: An environment/energy emulation and profiling platform for wireless sensor networks. in *Conference on Sensor and Ad Hoc Communications and Networks (SECON)*, 158–67. New York: IEEE Communications Society.

31. Brooks, D., and M. Martonosi. 2000. Value-based clock gating and operation packing: Dynamic strategies for improving processor power and performance. *ACM Trans. Computer Systems* 18 (2): 89–126.

32. Hamed, H., A. El-Atawy, and A.-S. Ehab. 2006. On dynamic optimization of packet matching in high-speed firewalls. *IEEE Journal on Selected Areas in Communications* 24 (10): 1817–30.

33. Hu, S., M. Valluri, and L. John. 2006. Effective management of multiple configurable units using dynamic optimization. *ACM Trans. Architecture and Code Optimization* 3 (4): 477–501.

34. Patel, S., and S. Lumetta. 2001. rePLay: A hardware framework for dynamic optimization. *IEEE Trans. Computers* 50 (6): 590–608.
35. Seong, C.-Y., and B. Widrow. 2001. Neural dynamic optimization for control systems. *IEEE Trans. Systems, Man, and Cybernetics* 31 (4): 482–89.
36. Zhang, C., F. Vahid, and R. Lysecky. 2004. A self-tuning cache architecture for embedded systems. *ACM Trans. Embedded Computing Systems* 3 (2): 407–25.
37. Dhodapkar, A., and J. Smith. 2002. Managing multi-configuration hardware via dynamic working set analysis. In *International Symposium on Computer Architecture (ISCA)*, 233–44. Washington, DC: IEEE Computer Society.
38. Kaxiras, S., Z. Hu, and M. Martonosi. 2001. Cache decay: Exploiting generational behavior to reduce cache leakage power. In *International Symposium on Computer Architecture (ISCA)*, 240–51. Washington, DC: IEEE Computer Society.
39. Douglis, F., P. Krishnan, and B. Bershad. 1995. Adaptive disk spindown policies for mobile computers. In *Symposium on Mobile and Location-Independent Computing (MLICS)*, 121–37. Berkeley, CA: USENIX Association.
40. Shnayder, V., M. Hempstead, B. Chen, G. Werner-Allen, and M. Welsh. 2004. Simulating the power consumption of large-scale sensor network applications. In *Conference on Embedded Networked Sensor Systems (SenSys)*, 188–200. New York: ACM Press.
41. Eriksson, J., F. Osterlind, N. Finne, A. Dunkels, N. Tsiftes, and T. Voigt. 2009. Accurate network-scale power profiling for sensor network simulators. In *Proceedings of the 6th European Conference on Wireless Sensor Networks (EWSN)*, 312–26. Berlin: Springer-Verlag.
42. Titzer, B., D. Lee, and J. Palsberg. 2005. Avrora: Scalable sensor network simulation with precise timing. In *International Conference on Information Processing in Sensor Networks (IPSN)*, 477–82. Piscataway, NJ: IEEE Press.
43. Landsiedel, O., K. Wehrle, S. Rieche, S. Gotz, and I. Petrak. 2005. Accurate prediction of power consumption in sensor networks. In *IEEE Workshop on Embedded Networked Sensors*, 37–44. Piscataway, NJ: IEEE Press.
44. Demaille, A., S. Peyronnet, and B. Sigoure. 2006. Modeling of sensor networks using XRM. In *International Symposium on Leveraging Applications of Formal Methods, Verification and Validation (ISoLA)*, 271–76. Piscataway, NJ: IEEE Press.
45. Kumar, U., A. Ranjan, V. Jalan, P. Mundra, and P. Ranjan. 2006. CENSE: A prototype for modular sensor network testbed. Paper presented at National Conference on Embedded Systems, Mumbai, India.
46. Handziski, V., A. Kopke, H. Karl, and A. Wolisz. 2003. A common wireless sensor network architecture. In *Sensornetze* (Technical Report TKN-03-012 of the Telecommunications Networks Group), 10–17. Berlin: Technical Universitat.
47. Sridharan, S., and S. Lysecky. 2008. A first step towards dynamic profiling of sensor-based systems. In *IEEE Conference on Sensor, Mesh and Ad Hoc Communications and Networks (SECON'08)*, 600–2. Piscataway, NJ: IEEE Press.
48. Shenoy, A., J. Hiner, S. Lysecky, R. Lysecky, and A. Gordon-Ross. 2010. Evaluation of dynamic profiling methodologies for optimization of sensor networks. *IEEE Embedded Systems Letters* 2 (1): 10–13.
49. Munir, A., and A. Gordon-Ross. 2009. An MDP-based application oriented optimal policy for wireless sensor networks. In *Conference on Hardware/Software Codesign and System Synthesis (CODES+ISSS)*, 183–92. New York: ACM Press.
50. Wang, X., J. Ma, S. Wang, and D. Bi. 2009. Distributed energy optimization for target tracking in wireless sensor networks. *IEEE Trans. Mobile Computing* 9 (1): 73–86.
51. Liu, L., X. Zhang, and H. Ma. 2009. Dynamic node collaboration for mobile target tracking in wireless camera sensor networks. In *IEEE INFOCOM'09*, 1188–96. Piscataway, NJ: IEEE Press.

52. Khanna, R., H. Liu, and H.-H. Chen. 2007. Dynamic optimization of secure mobile sensor networks: A genetic algorithm. In *IEEE International Conference on Communications (ICC'07)*, 3413–18. Piscataway, NJ: IEEE Press.

53. Min, R., T. Furrer, and A. Chandrakasan. 2000. Dynamic voltage scaling techniques for distributed microsensor networks. In *IEEE Workshop on VLSI (WVLSI'00)*, 43–46. Piscataway, NJ: IEEE Press.

54. Yuan, L., and G. Qu. 2002. Design space exploration for energy-efficient secure sensor network. In *IEEE International Conference on Application-Specific Systems, Architectures, and Processors (ASAP'02)*, 88–97. Piscataway, NJ: IEEE Press.

55. Lysecky, S., and F. Vahid. 2006. Automated application-specific tuning of parameterized sensor-based embedded system building blocks. In *Proceedings of International Conference on Ubiquitous Computing (UbiComp)*, 507–24. Berlin: Springer-Verlag.

56. R. Verma, R. 2008. Automated application specific sensor network node tuning for non-expert application experts. Masters thesis, Department of Electrical and Computer Engineering, University of Arizona.

57. Meier, A., M. Weise, J. Beutel, and L. Thiele. 2008. NoSE: Efficient initialization of wireless sensor networks. In *ACM Conference on Embedded Networked Sensor Systems (SenSys'08)*, 397–98. New York: ACM Press.

58. Hasegawa, M., T. Kawamura, N. Tran, G. Miyamoto, Y. Murata, H. Harada, and S. Kato. Decentralized optimization of wireless sensor network lifetime based on neural network dynamics. In *IEEE International Symposium on Personal, Indoor and Mobile Radio Communications (PIMRC'08)*, 1–5. Piscataway, NJ: IEEE Press.

59. Jurdak, R., P. Baldi, and C. Lopes. 2007. Adaptive low power listening for wireless sensor networks. *IEEE Trans. Mobile Computing* 6 (8): 988–1004.

60. Ma, D., J. Wang, M. Somasundaram, and Z. Hu. 2005. Design and optimization on dynamic power system for self-powered integrated wireless sensing nodes. In *IEEE International Symposium on Low Power Electronics and Design (ISLPED'05)*, 303–6. Piscataway, NJ: IEEE Press.

61. Ning, X., and C. Cassandras. 2008. Optimal dynamic sleep time control in wireless sensor networks. In *IEEE Conference on Decision and Control (CDC'08)*, 2332–37. Piscataway, NJ: IEEE Press.

62. Crossbow IRIS Datasheet. n.d. http://bullseye.xbow.com:81/Products/Product_pdf_files/Wireless_pdf/IRIS_Datasheet.pdf.

63. Atmel. 2012. ATmega1281 microcontroller with 256K bytes in-system programmable flash. http://www.atmel.com/images/doc2549.pdf.

64. Crossbow Technology. 2007. MTS/MDA Sensor Board Users Manual. http://bullseye.xbow.com:81/Support/Support_pdf_files/MTS-MDA_Series_Users_Manual.pdf.

65. Sensirion. 2010. Datasheet SHT1x (SHT10, SHT11, SHT15) Humidity and Temperature Sensor IC. http://www.sensirion.com/fileadmin/user_upload/customers/sensirion/Dokumente/Humidity/Sensirion_Humidity_SHT1x_Datasheet_V5.pdf.

66. Atmel. 2009. AVR Low Power 2.4 GHz Transceiver for ZigBee, IEEE 802.15.4, 6LoWPAN, RF4CE and ISM Applications. AT86RF230. http://www.atmel.com/Images/doc5131.pdf.

67. Fujita, M., Y. Matsunaga, and T. Kakuda. 1991. On variable ordering of binary decision diagrams for the application of multi-level logic synthesis. In Proceedings of European Conference on Design Automation (EDAC), 50–54. Piscataway, NJ: IEEE Press.

68. Intel Corp. n.d. Intel Xeon Processor E5430. http://ark.intel.com/Products/Spec/SLANU.

69. die.net. n.d. Linux Man Pages. http://linux.die.net/man/.

2 Stochastic Inference in Wireless Sensor Networks

Sahar Movaghati and Masoud Ardakani

CONTENTS

2.1 INFERENCE PROBLEMS

The continuing advances in sensor technology facilitate data acquisition and data processing with sensor devices. Due to these improvements, the new wireless sensor networks (WSNs) are able to gain more knowledge about the environment and perform more advanced tasks [1]. In the new applications of WSNs, handling and processing of the acquired data becomes very important. This means that the sensors' measurement data should be properly and efficiently processed in the network to achieve the most reliable and accurate inferences about the phenomenon under surveillance. In this sense, a WSN presents a distributed processing system that performs either information fusion [2], decision making, hypothesis testing [3], or distributed detection and estimation [4]. In general, all these concepts address theories and techniques to process and combine the measurements from multiple sensors for inferring/estimating a phenomenon pertaining to the observed system. In particular, in a large coverage area or in the case of temporally or spatially dynamic systems, a cooperative method can take advantage of the diversity and redundancy of multiple sensor measurements for more reliable and accurate results [1, 5].

The majority of the advanced tasks performed by WSNs, such as data fusion, decision making, hypothesis testing, detection, etc., can be interpreted in the form of a stochastic inference problem. In the stochastic inference, the object of interest as well as the sensors' measurement data are represented by random variables or processes. The measurements carry some information about the unknown object contaminated with a level of uncertainty. The uncertain/ambiguous relations among the variables of the problem are modeled via stochastic relations among the random variables. The result is called a stochastic model for the inference problem. Whether the goal of the network is to decide a hypothesis in the coverage area or estimate and track some time-varying object, the nature of the problem can be classified as a stochastic inference problem. In such problems, the sensors' measurements are called the "observable" variables. The goal of the network is to gain knowledge about some "unobservable" variable of interest. Based on the application, the unobservable variable could be the presence of fire in a forest or the location of an intruder vehicle. Viewing the WSN applications as a stochastic inference problem helps take advantage of the related literature in designing solution methods for WSN applications. In the following section, we explain the fundamentals of a stochastic inference problem.

2.1.1 GENERAL SETUP OF STOCHASTIC INFERENCE PROBLEMS

In recent decades, statistical approaches for solving many problems have become popular. There have been many works in statistics theory, methodology, and computation, such as Bayesian statistics [6]. As discussed previously, in a stochastic approach, all quantities involved in a problem, such as the observable variables (measurements) and unobservable variables (unknowns), are treated as random variables. For example, assume that the random variables $z_1, z_2, ..., z_L$ represent L measured values and $x_1, x_2, ..., x_M$ stand for M unknowns. The knowledge or belief associated with each random variable is described by a probability distribution function (PDF). For example, the PDF $p(x_m)$ explains some a priori knowledge about the unknown x_m. The greater the variance of this PDF, the greater is the uncertainty about the value of x_m. Furthermore, a joint probability distribution of the variables captures the relationship among the variables involved in the problem, i.e.,

$$p(x_1, x_2, ..., x_M, z_1, z_2, ..., z_L). \tag{2.1}$$

Furthermore, an a posteriori distribution is a conditional PDF such as

$$p(x_1, x_2, ..., x_M \mid z_1, z_2, ..., z_L), \tag{2.2}$$

which encapsulates all the information that could be achieved about the unknowns $x_1, x_2, ..., x_M$ knowing the measurements $z_1, z_2, ..., z_L$ [5, 7].

Normally, the amount of uncertainty in an a posteriori distribution is less than the uncertainty in an a priori distribution. The goal in a stochastic inference is to find the marginals of the a posteriori distributions. For example, the marginal a posteriori distribution of x_m, i.e.,

$$p(x_m \mid x_M, z_1, z_2, \ldots, z_L) = \int p(x_1, x_2, \ldots, x_M \mid z_1, z_2, \ldots, z_L) dx_1 \cdots dx_{m-1} dx_{m+1} dx_M, \quad (2.3)$$

is the a posteriori belief about the unknown x_m. If the PDF in Equation (2.3) is known, any desired estimate of x_m can be made regarding an estimation metric, for example, a maximum likelihood (ML) or a minimum mean squared error (MMSE) estimation.

This stochastic setup can be applied to many problems, ranging from single, time-invariant parameter estimation to multivariant time-varying processes. To solve these problems, i.e., finding the a posteriori marginal PDFs in Equation (2.3), many techniques in the literature have been developed, ranging from the Bayesian hypothesis testing to the Kalman filter (KF) [8], particle filtering [9–11], and graph-based methods such as belief propagation [12] and sum-product algorithms [13].

With the new advanced sensor and wireless technologies, a WSN can work as a distributed processing system for advanced applications. By viewing many of these applications as a stochastic inference problem, appropriate stochastic techniques can be deployed in the sensor network to solve the problem. However, due to many challenges and restrictions involved in WSNs, the conventional solutions are not applicable in many scenarios. Hence, there is a need to develop appropriate solutions for an inference problem in a WSN. In the next section, we describe some of these challenges.

2.1.2 CHALLENGES OF STOCHASTIC INFERENCE IN WSNs

The algorithms used to solve an inference problem in a WSN could be different from the conventional approaches. An important characteristic of these networks is that the measurements as well as the processing resources are distributed among the sensors. Therefore, any global estimation or inference should also be done in a distributed fashion. In some WSNs, a central fusion center (FC) with a powerful processing capability can be accessed by all the sensor nodes. In such networks, sensors transmit their measurements, i.e., $z_1, z_2, \ldots, z_L$, to the FC. Upon receiving the measurement data, a centralized solution can be run in the FC to find the a posteriori marginals using Equation (2.3). Following that, an appropriate decision or consensus can be done based on the a posteriori belief about the variables of interest, i.e., $x_1, x_2, \ldots, x_M$. However, this centralized approach suffers from many drawbacks that affect its performance. For example, in big multihop networks, it requires many retransmissions among the sensors, causing a big wireless communication load as well as increasing the response delay.

In some WSNs there is no FC; hence, the sensors should find the inference solution autonomously through cooperation. In such networks, a distributed algorithm must be designed to find a global solution for the inference problem by local processing in the sensors and relying on communications among the sensors in a neighborhood. In distributed algorithms, each sensor receives some messages from its neighbors and performs some local processing on the messages and then initiates a new message to its neighbors. The question is how to manage the local processing and data communication among the sensors to achieve a reliable estimate of the a posteriori PDFs in Equation (2.3). In Section 2.2, we benefit from a graphical tool to develop a distributed processing framework for general stochastic inference problems in WSNs.

Another inherent property of WSNs is the limited power resource in each sensor, i.e., the sensor's battery, which directly affects the sensor's functionalities. On the other hand, wireless communication itself is a major power-consuming job in the sensors [14]. Therefore, the sensors' capability for wireless communication is restrained by the limited power resources. These restrictions should be considered in the design of inference algorithms, which usually involve communication among the sensors.

Since transmitting high-precision data requires many bits, and therefore a lot of transmission power, data compression is a necessity in most WSNs. One approach is to quantize the quantities, e.g., z_l or x_m, to reduce the number of bits required for transmission between the sensors. Consequently, the framework and equations presented in Section 2.1.1 will be altered by a quantized version of these quantities instead of their real values. Clearly, the designed inference algorithm must be capable of handling these changes. For example, for the KF, which is the solution for a linear and Gaussian category of inference problems, the quantized version of the measurements necessitates some modifications to the KF equations [15, 16]. Data quantization has been addressed in many estimation algorithms for WSNs [17–24]; however, most of them are limited to a special category of inference problems, i.e., parameter estimation. In Section 2.3, a solution setup for the general inference problem is presented that could also incorporate data quantization.

2.2   FACTOR GRAPHS FOR SOLVING INFERENCE PROBLEMS

An appropriate tool that can be used to build a solution for a general stochastic inference problem is the factor graph (FG). Graphical models such as FGs have been used to model stochastic relationships of the variables involved in an inference problem. The FG takes advantage of the conditional independency among the variables to break down a global multivariate stochastic relation like the PDF in Equation (2.2) to smaller local functions. Therefore, a complicated global processing on the entire set of variables can be broken down to smaller processing modules that involve only some of the variables. Specifically, instead of calculating multivariable integrals to find the marginal beliefs using Equation (2.3), the FG representation of the stochastic problem helps derive a set of simpler processing rules that together achieve the same performance as the centralized solution. This setup enables a WSN to work as a powerful distributed processing entity for many complicated inference tasks. To understand how FGs help in the inference problems, the concept of the FG and the message-passing algorithm is briefly described here.

Assume a global function (a joint PDF) of a set of M variables, i.e., $\zeta(x_1,x_2,\ldots,x_M)$. factors into a product of several local functions, each having a subset of the M variables $\{x_1,x_2,\ldots,x_M\}$ as their arguments. This factorization can be written as

$$\zeta(x_1,x_2,\ldots,x_M) = \prod_j f_j(X_j), \qquad (2.4)$$

where X_j is a subset of $\{x_1,x_2,\ldots,x_M\}$, and $f_j(X_j)$ is a function having the elements of X_j as its arguments. In a stochastic inference problem, such factorization can be done based on the conditional stochastic independencies among the variables involved in the problem. For more information on this, refer to Chapter 2 in Jordan's book [25]. The probabilistic factorization of inference problems is explained more in Section 2.2.1 through some practical examples.

The factorization in Equation (2.4) can be shown using an FG, which contains a *variable node* for each variable x_i, and a *function node* for each local function f_j. There is an edge between the variable node x_i and the function node f_j if and only if x_i is an argument of f_j [13]. For example, if $\zeta(x_1,x_2,x_3,x_4,x_5)$ can be expressed as

$$\zeta(x_1,x_2,x_3,x_4,x_5) = f_1(x_1,x_2)f_2(x_3)f_3(x_2,x_3,x_4)f_4(x_4,x_5), \qquad (2.5)$$

then the FG corresponding to this factorization is the graph in Figure 2.1. In order to find all marginal functions of a global function, some calculations can be done on the FG according to a message-passing algorithm (MPA). An MPA is usually described by defining the *messages* that are passed along the edges of the FG and the *update rules* associated with the messages as they move through the nodes of the graph. A message sent along an edge is a function of the connected variable node to that edge and usually carries information about that variable. The message-update rules are procedures performed in the graph nodes to manipulate the incoming messages to the node in order to generate output messages from the node. A very famous MPA called the sum-product algorithm (SPA) is introduced in a work by Kschischang, Frey, and Loeliger [13]. By running the SPA on a cycle-free FG, once a message is passed in both directions on every edge, the marginal function of any variable can be found.

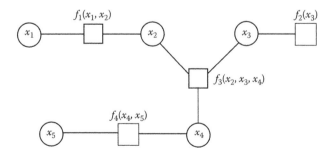

FIGURE 2.1 Factor-graph representation for the function ζ in Equation (2.5).

The SPA establishes the message update rules at the variable node x and the function node f. We represent the message going from a variable node to a function node as $\mu_{x \to f}(x)$ and a message from a function node to a variable node as $\mu_{f \to x}(x)$. Note that both messages are functions of the particular variable node involved in the transaction. Let $\aleph(x)$ indicate the set of neighbors of the variable node x and $\aleph(f)$ indicate the set of neighbors for the function node f. Then the update rules of the SPA are formulated as Update Rule 1

$$\mu_{x \to f}(x) = \prod_{\lambda \in \aleph(x) \backslash f} \mu_{\lambda \to x}(x) \tag{2.6}$$

and Update Rule 2

$$\mu_{f \to x}(x) = \sum_{\sim x} (f(X) \prod_{y \in \aleph(f) \backslash x} \mu_{y \to f}(y)), \tag{2.7}$$

where X is the set of the arguments of the local function f which will be the same as $\aleph(f)$. The notation $\aleph(x) \backslash f$ denotes the set of all the neighbors of x excluding the function node f. Also, $\sim x$ means that the sum (or the integral in the case of a continuous variable) is taken on all the variables except x.

In many inference problems, the global stochastic functions described in Section 2.1.1 can be broken down to a number of smaller stochastic relations. Once the factorization is realized, the inference problem can be represented by an FG, and consequently the SPA can be used to find the marginal a posteriori distributions.

2.2.1 Application of Factor Graphs in Inference Problems for WSNs

Graphical models have been frequently used for inference problems in WSNs [26–32]. Two of these applications are mentioned here to illuminate the benefit of graphical modeling in WSNs, focusing on FGs and the SPA.

2.2.1.1 Self-Tracking of Mobile Nodes

Robust self-tracking of mobile nodes in wireless networks is a critical enabler for numerous applications of high-definition location-aware networks. Wymeersch, Ferner, and Win [28] have used FG modeling to derive a distributed and cooperative algorithm for tracking the location of mobile nodes in a WSN.

Considering a WSN of N nodes, denote the position of node i at the discrete time step t to be $x_i^{(t)}$, and the aggregated positions of all the nodes at time t to be $\mathbf{x}^{(t)}$. Each node has a means of having some knowledge related to its own location (using an odometer, for example), as well as its neighbors' locations (for example, using some signaling between the nodes). Assume that all the measurements in the network up to time t are represented by $\mathbf{z}^{(1:t)}$, which can be decomposed into $\mathbf{z}_{\text{self}}^{(1:t)}$ and $\mathbf{z}_{\text{rel}}^{(1:t)}$. The vector $\mathbf{z}_{\text{self}}^{(1:t)}$ denotes all intranode measurements, while $\mathbf{z}_{\text{rel}}^{(1:t)}$ denotes all internode measurements. Due to the Markov chain nature of the problem and conditional independence, the a posteriori belief about the location of the nodes can be factorized as

$$p(\mathbf{x}^{(t)} \mid \mathbf{z}^{(1:t)}) \propto p(\mathbf{x}^{(0)}) \prod_{u=1}^{t} \left\{ p(\mathbf{x}^{(u)} \mid \mathbf{x}^{(u-1)}, \mathbf{z}_{\text{self}}^{(u)}) p(\mathbf{z}_{\text{rel}}^{(u)} \mid \mathbf{x}^{(u)}) \right\}.$$

(2.8)

In Equation (2.8), $p(\mathbf{x}^{(u)} \mid \mathbf{x}^{(u-1)}, \mathbf{z}_{\text{self}}^{(u)})$ and $p(\mathbf{z}_{\text{rel}}^{(u)} \mid \mathbf{x}^{(u)})$ can further be factorized as $\prod_{i=1}^{N} p(x_i^{(u)} \mid x_i^{(u-1)}, z_{i,\text{self}}^{(u)})$ and $\prod_{i=1}^{N} \prod_{j \in \mathcal{N}_i^{(u)}} p(z_{j \to i}^{(u)} \mid x_i^{(u)}, x_j^{(u)})$, respectively, where $\mathcal{N}_i^{(u)}$ is the set of nodes from which node i can receive transmissions at time t, i.e., the neighbors of node i, and $z_{j \to i}^{(u)}$ indicates the measurement value between node i and j. This factorization leads to an FG representation of the problem as illustrated in Figure 2.2. The sum-product algorithm is then run to find the marginal a posteriori beliefs about the locations.

2.2.1.2 Link Loss Monitoring

The dynamic nature of wireless environments makes it desirable to monitor link loss rates in WSNs. In a WSN, the data-collecting node can infer link loss rates on all links in the network by exploiting whether packets from various sensors are received. Mao et al. [29] present an algorithm for link loss monitoring based on FG methodology. The algorithm iteratively updates the estimates of link losses upon receiving (or detecting the loss of) recently sent packets by the sensors.

In this problem, Mao et al. [29] have assumed that all the nodes in the network (terminals) constantly send packets destined to a single center node. The packets go through single or multihop paths, i.e., each "path" consists of a number of "links." The intermediate nodes aggregate all the received packets into a larger packet and send it to their parent node. If a packet is not received by the center node within a specific time frame, a packet loss is suggested to have occurred on one of the links along the

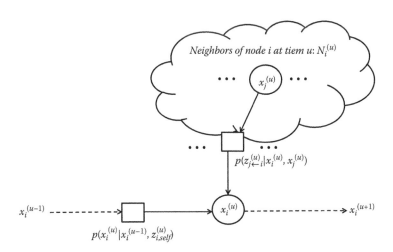

FIGURE 2.2 Factor-graph representation for the self-tracking problem. (*Source*: [28].)

path. Based on the successive observations on whether packets from each terminal have arrived, the center node can infer the link loss rates on all links in the network.

In the problem setup, $E = \{e_1,...,e_{|E|}\}$ denotes the set of all links, and $W = \{w_1,...,w_{|W|}\}$ denotes all the paths in the network. The state of each link $e_i \in E$ is represented by a Bernoulli random variable x_i, taking values either 0 with probability α_i and 1 with probability $1 - \alpha_i$, representing a "bad" or "good" link, respectively. The state of a path $w_j \in W$ is $z_j = \oplus X_j$, where X_j is the set of states of those links contained in the path w_j. Also, $\oplus$ denotes the logic *AND* operator on all the elements of X_j. For example, in Figure 2.3, $X_1 = \{x_1,x_2\}$ and $X_2 = \{x_2,x_3\}$.

The objective is to estimate the link loss rates α_i at the end of each time window during which n batches of synchronized packets have been sent. Therefore, it is desired to find the marginals of the a posteriori $p\left(\alpha_1,...,\alpha_{|E|},x_1^{(1:n)},...,x_{|E|}^{(1:n)} \mid z_1^{(1:n)},...,z_{|W|}^{(1:n)}\right)$, where the super index $(1:n)$ indicates the ensemble of all n batches. Using the relative independence in this problem the a posteriori PDF can be factorized as (for more information about this factorization please see Mao et al. [29])

$$p\left(\alpha_1,...,\alpha_{|E|},x_1^{(1:n)},...,x_{|E|}^{(1:n)} \mid z_1^{(1:n)},...,z_{|W|}^{(1:n)}\right) \propto \prod_{k=1}^{n}\left\{\prod_{i=1}^{|E|} p(x_i^{(k)} \mid \alpha_i)\prod_{j=1}^{|W|} p(z_j^{(k)} \mid X_j^{(k)})\right\}.$$

(2.9)

In Equation (2.9), $p(x_i^{(k)} \mid \alpha_i)$ is the probability mass function (PMF) of a Bernoulli random variable parameterized by α_i, and it can be written as

$$p(x_i^{(k)} \mid \alpha_i) = \begin{cases} \alpha_i & x_i^{(k)} = 1 \\ 1-\alpha_i & x_i^{(k)} = 0 \end{cases}.$$

(2.10)

Also, $p(z_j^{(k)} \mid X_j^{(k)})$ is simply $\delta(z_j^{(k)} - \oplus X_j^{(k)})$, where $\delta(\cdot)$ is the discrete Dirac delta function, with value 1 if $z_j^{(k)} = \oplus X_j^{(k)}$ and 0 otherwise. The FG representation of

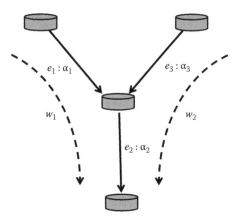

FIGURE 2.3   A sample WSN with link loss monitoring.

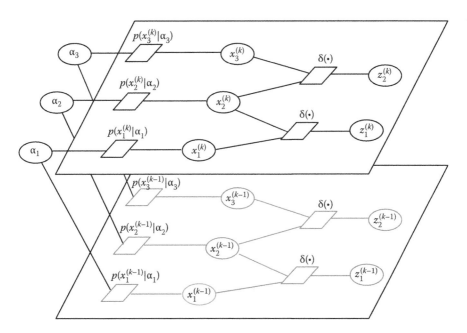

FIGURE 2.4 Factor-graph representation for the link loss monitoring problem. (*Source*: [29].)

Equation (2.9) is shown in Figure 2.4. An appropriate sum-product algorithm is run based on the FG to find the marginal beliefs for each α_i, from which the link loss rate can be approximated.

To find the desired marginal a posteriori PDF in the discussed problems, the SPA is run on the FG. In SPA for these problems, the messages are either discrete quantities or simple functions. Therefore, the message-update rules at the nodes of the graph are easy to calculate. This simple SPA is only realizable when the variables of the problem are discrete or the stochastic relations among the variables are simple functions, such as Gaussian. In a general scenario, however (i.e., continuous variables and arbitrary stochastic functions), the messages on the FG are arbitrary continuous functions, making the message update rules difficult to calculate. For such general scenarios, a nonparametric MPA is required to be run on the FG. In the next section, a general nonparametric solution for finding the marginal a posteriori beliefs on an FG is proposed.

2.3 PARTICLE-BASED MESSAGE-PASSING ALGORITHM

In our proposed MPA, the messages, which are generally continuous PDFs, are represented by a set of N random "particles" and their "importance weights," i.e., $\mu(x) = \{x^i, w^i; 1 \leq i \leq N\}$. Therefore, we call this method the particle-based message passing algorithm (PB-MPA). Our goal is to formulate update rules, Equations (2.6) and (2.7), at variable nodes and function nodes of an FG for such messages. In other words, the question is how to update x^i and w^i according to the SPA. This formulation

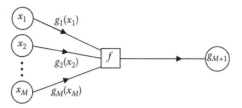

FIGURE 2.5   A function node in a factor graph.

is performed in two parts. In Section 2.3.1, we formulate the message update rule at a function node of the FG. In Section 2.3.2, we develop the message update rule at a variable node of the FG.

2.3.1 MESSAGE PROCESSING AT A FUNCTION NODE

In this step, we use the Monte Carlo integration concept [9] to approximate the integral at a function node. Assume that there are $M + 1$ variable nodes connected to a function node f, as seen in Figure 2.5. Recalling Equation (2.7), we rewrite the outgoing message from the function node f to the $(M + 1)$th variable node x_{M+1} as

$$\mu_{f \to x_{M+1}}(x_{M+1}) = h(x_{M+1}) = \int f(x_1,...,x_M,x_{M+1})g_1(x_1)...g_M(x_M)dx_1...dx_M. \quad (2.11)$$

In Equation (2.11), $x_1,...,x_M$ are M variable nodes connected to the function node f, and $g_1(x_1),...,g_M(x_M)$ are their corresponding messages toward f. Note that in statistical inference problems, a message coming out of a variable node, e.g., $g_m(x_m); 1 \le m \le M$, is a conditional PDF. Defining the vector $X = [x_1,...,x_M]^T$, the M-dimensional integral of Equation (2.11) can be written as a vector integral

$$h(x_{M+1}) = \int f(x_{M+1},X)G(X)dX, \quad (2.12)$$

where $G(X)$ can be viewed as the joint distribution of $x_1,...,x_M$, which according to Equation (2.11) is equal to $\prod_{m=1}^{m=M} g_m(x_m)$. Using the Monte Carlo integration method [9], $h(x_{M+1})$ can be approximated as the sample mean of $f(x_{M+1},X)$, i.e.,

$$h(x_{M+1}) = E_X\{f(x_{M+1},X)\}$$

$$\simeq \frac{1}{N}\sum_{i=1}^{N} f(x_{M+1},X^i), \quad \text{where} \quad X^i \sim G(X). \quad (2.13)$$

This equation is a function of x_{M+1} and can be regarded as the outgoing message from the function node f, i.e., $\mu_{f \to xM+1}(x_{M+1})$. However, $h(x_{M+1})$ is a continuous function that should be represented as a particle-based message for the next stage of our PB-MPA. Therefore, we need to build the particle representation of $h(x_{M+1})$

as $\{x_{M+1}^i, w_{M+1}^i; 1 \leq i \leq N\}$. The best way to construct a particle representation is to directly draw N i.i.d. (independent and identically distributed) samples from $h(x_{M+1})$. However, it is not straightforward to sample from $h(x_{M+1})$ because it is a sum of N continuous functions. Hence, we use the importance sampling method [9] to construct the particle representation of the message $\mu_{f \to x_{M+1}}(x_{M+1})$. This means that, instead of drawing samples from $h(x_{M+1})$, we first take samples from another PDF called the importance function, and then we compensate by assigning an importance weight to each sample. The importance distribution function $q(x_{M+1})$ can be any positive function with the same support as $h(x_{M+1})$. The distribution $q(x_{M+1})$ is chosen such that it is straightforward to sample from. Then we draw N samples from $q(x_{M+1})$ and assign an importance weight to each sample, w_{M+1}^i. Thus,

$$\mu_{f \to x_{M+1}}(x_{M+1}) = h(x_{M+1}) \simeq \{x_{M+1}^i, w_{M+1}^i\} \quad 1 \leq i \leq N, \tag{2.14}$$

where

$$x_{M+1}^i \sim q(x_{M+1}) \qquad w_{M+1}^i \propto \frac{h(x_{M+1}^i)}{q(x_{M+1}^i)}.$$

The importance weights should be normalized so that their sum is equal to 1. Here, without loss of generality, we choose the importance distribution function to be $f(x_{M+1}, X^1)$. If it is not possible to sample from the function $f(x_{M+1}, X^1)$, one must choose another appropriate function as the importance density function and adjust the importance weights accordingly. Thus, to determine the ith particle and its corresponding weight, we set $q(x_{M+1}) = f(x_{M+1}, X^1)$. By drawing the ith particle x_{M+1}^i from $f(x_{M+1}, X^1)$, the ith importance weight will be proportional to the ratio of $h(x_{M+1})$ to $f(x_{M+1}, X^1)$ evaluated at x_{M+1}^i. Therefore, the particle representation of $\mu_{f \to x_{M+1}}(x_{M+1})$ is

$$\mu_{f \to x_{M+1}}(x_{M+1}) \simeq \{x_{M+1}^i, w_{M+1}^i\} \qquad 1 \leq i \leq N, \tag{2.15}$$

where

$$x_{M+1}^i \sim f(x_{M+1}, X^1) \tag{2.16}$$

and

$$w_{M+1}^i \propto \frac{\frac{1}{N} \sum_{j=1}^N f(x_{M+1}^i, X^j)}{f(x_{M+1}^i, X^1)}. \tag{2.17}$$

Note that any of the functions $f(x_{M+1}, X^i)$, where $1 \leq i \leq N$, can be chosen to be the importance function in Equation (2.15).

Based on the explained procedure formulated in Equations (2.15)–(2.17), we need N (weighted) samples of vector X. As discussed previously, X can be viewed as an ordered M-tuple with independent entries, i.e., $x_1,\ldots,x_M$, distributed according to $g(1),\ldots,g(M)$, respectively. Thus, a sample of X can easily be constructed by putting together the samples of the messages coming from $x_1,\ldots,x_m$. In our setup, however, samples are weighted. If all samples are equally weighted (i.e., all messages are represented by equally weighted particles), taking one particle from each message at a time, we can build a particle for X. When particles have unequal weights, we first use a resampling procedure to extract N equally weighted particles, and then continue with the normal procedure. This resampling algorithm draws i.i.d. samples from a set of particles with nonequal weights.

2.3.2 MESSAGE PROCESSING AT A VARIABLE NODE

At a variable node, we use the method developed by Sudderth et al. [33]. Assume that $L + 1$ function nodes, $f_1,\ldots,f_{L+1}$, are connected to a variable node x, as shown in Figure 2.6. The lth incoming messages is presented as $h_l(x) = \{x_l^i, w_l^i\}, 1 \leq i \leq N$. We now represent $h_l(x)$ with a Gaussian mixture of N weighted Gaussian kernels, each with mean x_l^i, variance σ_l^2, and weight w_l^i, $1 \leq i \leq N$. Notice that for simplicity, we have chosen all the kernel variances of the lth Gaussian mixture to be equal. A *rule of thumb* is to set σ_l^2 equal to the weighted variance of the samples of the lth message divided by $N^{1/6}$ [34]. Hence,

$$\mu_{f_l \to x}(x) = h_l(x) = \sum_{i=1}^{N} w_l^i \mathcal{N}(x; x_l^i, \sigma_l^2), \qquad (2.18)$$

where $\mathcal{N}(x; x^i, \sigma_l^2)$ is a Gaussian function of the variable x, with mean x_l^i and variance σ_l^2. According to Equation (2.6), the outgoing message $\mu_{x \to f_{L+1}}(x)$ will be written as

$$\mu_{x \to f_{L+1}}(x) = g(x) = \prod_{l=1}^{L} \sum_{i=1}^{N} w_l^i \mathcal{N}(x; x_l^i, \sigma_l^2). \qquad (2.19)$$

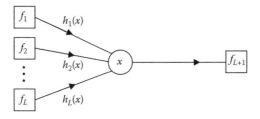

FIGURE 2.6 A variable node in a factor graph.

This equation is a product of L Gaussian mixtures, each having N components. Since the product of any number of Gaussian functions is proportional to a Gaussian function, by expanding Equation (2.19), we have N^L Gaussian components. Each component has a new mean, variance, and weight. Our goal is to find N weighted particles from Equation (2.19) to represent $\mu_{x \to f_{L+1}}(x) = g(x)$. Obviously, working with N^L components is not efficient. Thus, similar to Sudderth et al. [33], we use the Gibbs sampler to obtain asymptotically unbiased samples from the product of L Gaussian mixtures.

2.3.3 APPLICATION OF PB-MPA IN MANEUVERING TARGET TRACKING WITH WSNs

To elaborate the application of PB-MPA, two inference problems in WSNs are discussed. The stochastic model of the problem is presented by an FG. Since each of these problems involves some nonlinearity in the stochastic relations, the nonparametric MPA proposed in Section 2.3, i.e., the PB-MPA, is used to solve the inference problem.

The first application is a target-tracking problem using a WSN. The target is a moving object that is monitored at each time interval k by $S_k \geq 3$ number of sensors. We assume that the sensors know their own location as well as all the other sensors in the surveillance area, which is a common assumption in tracking problems (e.g., see [16, 35, 36]). It is further assumed that the data transmission is not subject to communication errors. Sensors quantize their measurements using a Q-level quantizer and transmit the quantized data bits over the wireless channel to their neighboring nodes. Each node receives some quantized measurement data from its neighbors. Then, each sensor runs a local tracking algorithm using its own measurement as well as the received data from its neighbors to estimate the location of the target at time step k. At the end of time step k, each sensor in the target area[*] has a local estimation of the target's state. Therefore, each sensor in the target area can identify the closest sensor to the target based on its own estimation of the location of the target and the sensors' location information. The sensor that finds itself the closest sensor to the target broadcasts its estimation so that the sensors involved in the next step of the estimation process have some knowledge about the target's state at the previous step. The closest sensor broadcasts its estimation because it is more likely to have the best estimation of the target's state.

To develop the target tracking algorithm, first we have to model the problem. The unknown quantity to be estimated is the location of the target in the xy-plane at each time step k, i.e., $P_k = [x_k \, y_k]^T$. For better tracking results, a random Markov chain process model is often assumed for the target dynamics. From the various target models proposed in the literature [37], a maneuvering-target model has been chosen here based on suitability. In the maneuvering-target model, a velocity vector is associated with the target as $V_k = [vx_k \, vy_k]^T$. The target also has an acceleration, $A_k = [ax_k \, ay_k]^T$,

---

[*] A sensor is in the target area if it can have a measurement from the target.

whose elements take a discrete value from the set $\{0, -g, +g\}$. Assuming a first-order Markov process, the target dynamics are explained as

$$X_k = \begin{bmatrix} P_k \\ V_k \end{bmatrix} = \begin{bmatrix} x_k \\ y_k \\ vx_k \\ vy_k \end{bmatrix} = F \begin{bmatrix} x_{k-1} \\ y_{k-1} \\ vx_{k-1} \\ vy_{k-1} \end{bmatrix} + G \begin{bmatrix} ax \\ ay \end{bmatrix} + G \begin{bmatrix} u_x \\ u_y \end{bmatrix}, \quad (2.20)$$

where

$$F = \begin{bmatrix} 1 & 0 & t_s & 0 \\ 0 & 1 & 0 & t_s \\ 0 & 0 & 1 & 0 \\ 0 & 0 & 0 & 1 \end{bmatrix} \quad G = \begin{bmatrix} t_s^2/2 & 0 \\ 0 & t_s^2/2 \\ t_s & 0 \\ 0 & t_s \end{bmatrix},$$

where t_s is the step size, and $[u_x \ u_y]^T$ is the process noise. Each of the two components of the acceleration vector at time step k is either 0, $-g$, or $+g$ with probabilities modeled as a random Markov jump with the initial 3×1 probability vector Pr_0 and the 3×3 transition matrix Tr.

The next step is to define a measurement model that relates the sensors' measurements to the target states. In this tracking scenario, the sensors are only able to measure their distance to the target. Having only distance measurements, at least three measurement values are required to infer a two-dimensional (2-D) location. Let $[n_x \ n_y]^T$ define the location of a sensor in a 2-D plane, then the measurement model is

$$z_k = \sqrt{(x_k - n_x)^2 + (y_k - n_y)^2} + v_k, \quad (2.21)$$

where v_k is the measurement noise, which stands for the error in measuring the distance. Notice that at time step k, there are S_k sensors measuring the distance of the target. Therefore, there are S_k measurement equations similar to Equation (2.21) with different n_x and n_y values. For the ease of notation, we have ignored the sensor index in Equation (2.21) in parts of the subsequent discussions.

Each sensor estimates its distance to the target at each time step k and sends it to its one-hop neighbors. To achieve power efficiency in WSNs, the high-precision sensors' measurements are quantized prior to transmission on the channel. One way to do this is to quantize the absolute z_k. This is not very efficient because z_k ranges from zero to large numbers, and hence the quantization requires many bits to convey enough information. A better way is to quantize a relative value instead of the absolute value. In this method, we quantize the normalized innovation data, which is defined as

$$I_k = \frac{z_k - \hat{z}_{k-1}}{N_f}, \quad (2.22)$$

where

$$\hat{z}_{k-1} = \sqrt{(\hat{x}_{k-1} - n_x)^2 + (\hat{y}_{k-1} - n_y)^2}.$$

Also, N_f is a normalization factor that estimates the maximum value of $z_k - \hat{z}_{k-1}$ and can be found to be

$$N_f = t_s V_{\max} + \frac{t_s^2}{2} g + 5 \frac{t_s^2}{2} \sigma_{vk} \qquad (2.23)$$

where σ_{vk} is the measurement noise standard deviation, and $V_{\max}$ is a rough estimation of the maximum velocity of the target. Having the measurement at time k, each sensor calculates the innovation value using Equation (2.22). According to the definition of N_f, almost always $I_k \in [-1, +1]$. Thus, to quantize I_k, we use a Q-level quantizer that quantizes the range $[-1, +1]$ to Q levels. This quantized value is referred to as q_k. In the rare cases that I_k is outside the range $[-1, +1]$, it is truncated to -1 or $+1$ accordingly.

The objective is to infer the state of the target, X_k, at each time step k from the quantized innovation data of $S_k \geq 3$ sensors. Referring to Section 2.1.1, one should find the marginal a posteriori PDF of X_k, which is a marginal function of the joint a posteriori PDF $\mathbf{f}(X_k, z_k, q_k \mid X_{k-1}, \hat{X}_{k-1}, A_{k-1})$. Note that at time step k we already have the a posteriori PDF of X_{k-1}, from which we estimate $\hat{X}_{k-1}$. The next step is to factorize this joint PDF based on the conditional independencies. This factorization can be done based on Equations (2.20)–(2.22), which results to

$$\mathbf{f}(X_k, z_k, q_k \mid X_{k-1}, \hat{X}_{k-1}, A_{k-1}) = \mathbf{f}(X_k \mid X_{k-1}, A_{k-1}) \mathbf{f}(z_k \mid X_k) \mathbf{f}(q_k \mid z_k, \hat{X}_{k-1}). \qquad (2.24)$$

The FG associated with the factorization in Equation (2.24) is sketched in Figure 2.7. Because of the nonlinear relation in Equation (2.21) and the quantization used to generate q_k, the third term in the factorization in Equation (2.24) is not a simple function. Also, the arbitrary measurement noise PDF can make the second term a non-Gaussian function. Therefore, the nonparametric solution of PB-MPA should be applied according to the update rules explained in Sections 2.3.1 and 2.3.2.

At step k, the message passing starts from X_{k-1} (node v_4) and $\hat{X}_{k-1}$ (node v_1). From step $k - 1$, the a posteriori PDF of X_{k-1} is approximated by N particles and their importance weights in the form of $X_{k-1} \simeq \{x^i, w^i; 1 \leq i \leq N\}$. Therefore, we can also calculate $\hat{X}_{k-1}$ based on an estimation criterion, e.g., ML or MMSE. The message going out from v_2 is q_k, known from the quantized innovation value of the sensor. Knowing the messages from the nodes v_1, v_2, and v_4, as well as all the function nodes, the PB-MPA can be used to find the marginal distribution of X_k at v_5. The only unusual calculation is the message passing on the upper branch of the FG of Figure 2.7, i.e., nodes v_6, f_4, and v_7. This branch represents the progress of target acceleration through time. As mentioned previously, the target's acceleration is

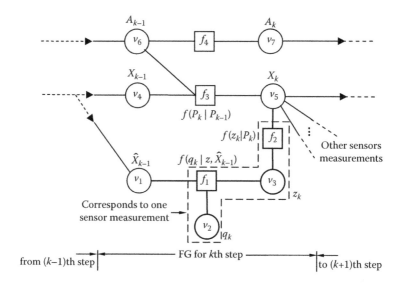

FIGURE 2.7 The factor graph of the tracking problem in a WSN.

modeled as a Markov jump. Therefore, one can easily find the PMF of the i.i.d. random variables ax_k and ay_k by propagating the PMF of ax_{k-1} and ay_{k-1} using the transition matrix, Tr. Therefore, the calculation at the function node f_4 would be $Pr_k = Pr_{k-1} \cdot Tr$. Let Pr_k have a general form of $[p_{1k}, p_{2k}, p_{3k}]$, where $p_{1k} + p_{2k} + p_{3k} = 1$; then the PMF of either components of the acceleration is of the form

$$f_k(a) = p_{1_k}\delta(a) + p_{2_k}\delta(a - g) + p_{3_k}\delta(a + g), \tag{2.25}$$

where $\delta(a)$ indicates the discreet delta function. The message going from variable node v_6 to the function node f_3 is a set of N 2-D vectors, $A_{k-1}^i, 1 \le i \le N$, randomly drawn according to the PMF $f_k(a)$.

The PB-MPA was tested in a simulated target tracking scenario in a WSN. In our example, $S_k = 5$ sensors is located inside an area of size 20×20, and every sensor can hear the other sensors' transmissions. At each time interval, all five sensors take a noisy measurement of the target's distance to themselves and broadcast the quantized version. The target is moving according to Equation (2.20), where $[u_x u_y]^T$ is a zero-mean Gaussian noise with covariance matrix $\Sigma_d^2 = 0.02 \times \mathcal{I}$, where $\mathcal{I}$ represents the 2×2 identity matrix. Also, we assume that $t_s = 1$, and the acceleration vector is a Markov jump process with $g = 0.1$ and

$$Pr_0 = \begin{bmatrix} 0.7 & 0.15 & 0.15 \end{bmatrix} \quad \text{and} \quad Tr = \begin{bmatrix} 0.6 & 0.2 & 0.2 \\ 0.4 & 0.5 & 0.1 \\ 0.4 & 0.1 & 0.5 \end{bmatrix}.$$

The measurement noise in Equation (2.21) is also assumed to be zero-mean Gaussian with variance σ_m^2. The algorithm, however, works with any other noise distribution. In the simulations, we have chosen σ_m proportional to the distance between the sensor and the target. Therefore, a sensor farther from the target has a noisier measurement of its distance than a closer sensor to the target. For a distance z, a signal-to-noise ratio is defined as SNR $= \frac{z^2}{\sigma_m^2}$. Figure 2.8 depicts a sample outcome of the target tracking compared to two other algorithms, i.e., a Gaussian MPA and the extended Kalman filter (EKF) [9].

2.3.4   APPLICATION OF PB-MPA IN MULTITARGET TRACKING WITH WSNS

The second inference problem is the more complicated scenario of multitarget tracking, which involves the data-association problem [38–40]. In this problem, there are a number of targets being tracked by a network of wireless sensors. At each time step, each sensor has some observations related to some of the targets. The challenging task is how to associate the measurements with the targets and track all targets simultaneously. Most of the existing solutions treat the data-association task first, and then apply conventional single-target tracking algorithms. Using the PB-MPA, these two steps can be solved in a conjunctive fashion.

Since a 2-D tracking scenario has been discussed in Section 2.3.3, to simplify the equations in this example we assume a one-dimensional (1-D) trajectory for each target. We consider two targets being tracked by a single sensor. The extension of the algorithm to multiple sensors is straightforward and identical to the approach of Section 2.3.3.

Assume that the location of targets 1 and 2, at time step k, are denoted by T_k^1 and T_k^2, respectively. At each time step, the sensor has two distance measurements from the targets, i.e., z^A and z^B, but it does not know which measurement is related to which target. Having the two measurement values, there are two association hypotheses: H_0:z^A is related to the first target, and z^B is therefore related to the

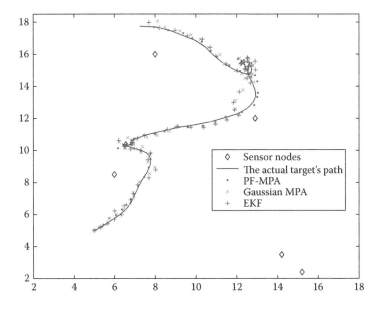

FIGURE 2.8   The result of tracking algorithm for $Q = 4$, $N = 40$, SNR = 50 dB.

second target; H_1:z^A is related to the second target, and z^B is therefore related to the first target. We define a hidden variable r that determines the association hypothesis. Therefore, r is a Bernoulli random variable that takes either the value 0, indicating H_0, or 1, indicating H_1. To finish the stochastic model setup, we also define z_k^1 and z_k^2 to be the expected measurement values for targets 1 and 2, respectively.

We can factorize the joint a posteriori PDF $p(T_k^1, T_k^2, z_k^1, z_k^2, r \mid T_{k-1}^1, T_{k-1}^2, z^A, z^B)$ as

$$p\left(T_k^1, T_k^2, z_k^1, z_k^2, r \mid T_{k-1}^1, T_{k-1}^2, z^A, z^B\right)$$

$$= p(r \mid z_k^1, z_k^2, z^A, z^B) p(z_k^1 \mid T_k^1) p(z_k^2 \mid T_k^2) p(T_k^1 \mid T_{k-1}^1) p(T_k^2 \mid T_{k-1}^2). \quad (2.26)$$

The associated FG is shown in Figure 2.9. In the factorization in Equation (2.26), the factors $p(T_k^1 \mid T_{k-1}^1)$ and $p(T_k^2 \mid T_{k-1}^2)$ can easily be derived based on the dynamic model that we consider for targets 1 and 2, respectively, similar to the approach in Section 2.3.3. Also, the factors $p(z_k^1 \mid T_k^1)$ and $p(z_k^2 \mid T_k^2)$ are decided based on

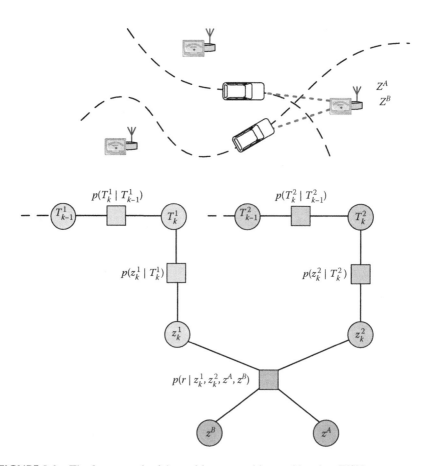

FIGURE 2.9 The factor graph of the multitarget tracking problem in a WSN.

the measurement models assumed for each target. The only remaining factor is $p(r \mid z_k^1, z_k^2, z^A, z^B)$, which basically is the probability of the hypothesis H_0 or H_1, knowing the values of the expected measurements z_k^1 and z_k^2, and the actual measurements z^A and z^B. An appropriate probability function must be chosen to represent $p(r \mid z_k^1, z_k^2, z^A, z^B)$, e.g.,

$$
p(r \mid z_1(k), z_2(k), z_A, z_B) =
\begin{cases}
\tanh\left(K \dfrac{\mid z_k^1 - z^B \parallel z_k^2 - z^A \mid}{\mid z_k^1 - z^A \parallel z_k^2 - z^B \mid} \right); & r = 0 \\[4mm]
1 - \tanh\left(K \dfrac{\mid z_k^1 - z^B \parallel z_k^2 - z^A \mid}{\mid z_k^1 - z^A \parallel z_k^2 - z^B \mid} \right); & r = 1
\end{cases}
,
$$

$$(2.27)$$

where K is a constant that is chosen to be 0.55 in our simulations. The PB-MPA can now run on the FG of Figure 2.9 to find the marginal a posteriori PDFs of $T_1(k)$ and $T_2(k)$. A sample result is shown in Figure 2.10. In this example, both targets' models are a 1-D Markov chain with model noise variance of 0.0001. The measurement noise variance is 0.0001 for the top graph and 0.0004 for the bottom graph.

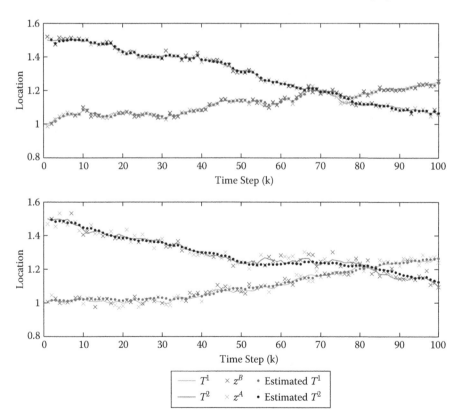

FIGURE 2.10   The sample results of multitarget tracking using PB-MPA.

REFERENCES

1. Ahmed, M. 2002. Decentralized information processing in wireless peer to peer networks. Ph.D. diss., UCLA, EE Dept.
2. Nakamura, E. F., A. A. F. Loureiro, and A. C. Frery. 2007. Information fusion for wireless sensor networks: Methods, models, and classifications. *ACM Comput. Surv.* 39 (3): 9.
3. Chen, B., L. Tong, and P. Varshney. 2006. Channel-aware distributed detection in wireless sensor networks. *IEEE Signal Process. Mag.* 23 (4): 16–26.
4. Varshney, P. K. 1996. *Distributed detection and data fusion.* New York: Springer-Verlag.
5. Mitchell, H. 2007. *Multi-sensor data fusion: An introduction.* New York: Springer.
6. Dey, D. K., and C. R. Rao, eds. 2005. *Bayesian thinking modeling and computation.* Vol. 25 of *Handbook of statistics.* North Holland: Elsevier.
7. Fearnhead, P. 1998. Sequential Monte Carlo methods in filter theory. Ph.D. diss., University of Oxford.
8. Kalman, R. 1960. A new approach to linear filtering and prediction problems. *Trans. of the ASME, Journal of Basic Eng.* 82:35–45.
9. Ristic, B., S. Arulampalam, and N. Gordon. 2004. *Beyond the Kalman filter: Particle filters for tracking applications.* London: Artech House.
10. Arulampalam, M., S. Maskell, N. Gordon, and T. Clapp. 2002. A tutorial on particle filters for online nonlinear/non-Gaussian Bayesian tracking. *IEEE Trans. Signal Process.* 50 (2): 174–188.
11. Doucet, A., S. Godsill, and C. Andrieu. 2000. On sequential Monte Carlo sampling methods for Bayesian filtering. *Statistics and Computing* 10:197–208.
12. Yedidia, J. S., W. T. Freeman, and Y. Weiss. 2002. Understanding belief propagation and its generalizations. Tech. Rep. TR-2001-22, Mitsubishi Elect. Res. Lab.
13. Kschischang, F., B. Frey, and H.-A. Loeliger. 2001. Factor graphs and the sum-product algorithm. *IEEE Trans. Inf. Theory* 47 (2): 498–519.
14. Willig, A. 2006. Wireless sensor networks: Concept, challenges and approaches. *Elektrotechnik und Informationstechnik* 123 (6): 224–231.
15. Msechu, E., S. Roumeliotis, A. Ribeiro, and G. Giannakis. 2008. Decentralized quantized Kalman filtering with scalable communication cost. *IEEE Trans. Signal Process.* 56 (8): 3727–41.
16. Ribeiro, A., G. Giannakis, and S. Roumeliotis. 2006. SOI-KF: Distributed Kalman filtering with low-cost communications using the sign of innovations. *IEEE Trans. Signal Process.* 54 (12): 4782–95.
17. Li, J., and G. AlRegib. 2009. Distributed estimation in energy-constrained wireless sensor networks. *IEEE Trans. Signal Process.* 57 (10): 3746–58.
18. Xiao, J.-J., A. Ribeiro, Z.-Q. Luo, and G. Giannakis. 2006. Distributed compression-estimation using wireless sensor networks. *IEEE Signal Process. Mag.* 23 (4): 27–41.
19. Luo, Z.-Q., and J.-J. Xiao. 2005. Universal decentralized estimation in a bandwidth constrained sensor network. *Proc. Int. Conf. Acoust., Speech, Signal Process.* 4:829–32.
20. Xiao, J., S. Cui, Z. Luo, and A. Goldsmith. 2006. Power scheduling of universal decentralized estimation in sensor networks. *IEEE Trans. Signal Process.* 54 (2): 413–22.
21. Xiao, J.-J., and Z.-Q. Luo. 2005. Decentralized estimation in an inhomogeneous sensing environment. *IEEE Trans. Inf. Theory* 51 (10): 3564–75.
22. Fang, J., and H. Li. 2008. Distributed adaptive quantization for wireless sensor networks: From delta modulation to maximum likelihood. *IEEE Trans. Signal Process.* 56 (10): 5246–57.
23. Ribeiro, A., and G. Giannakis. 2006. Bandwidth-constrained distributed estimation for wireless sensor networks-part I: Gaussian case. *IEEE Trans. Signal Process.* 54 (3): 1131–43.

24. Movaghati, S., and M. Ardakani. 2011. Energy-efficient quantization for parameter estimation in inhomogeneous WSNs. In *Global Telecommunications Conference (GLOBECOM 2011)*, 1–5. Piscataway, NJ: IEEE.

25. Jordan, M. I. 2003. Graphical models in *The Handbook of Brain Theory and Neural Networks, Second Edition.*, M. A. Arbib, ed. Cambridge: MIT Press.

26. Makarenko, A., A. Brooks, T. Kaupp, H. Durrant-Whyte, and F. Dellaert. 2009. Decentralised data fusion: A graphical model approach. In *Information Fusion (FUSION '09), 12th International Conference on*, 545–54. Piscataway, NJ: IEEE.

27. Moura, J., J. Lu, and M. Kleiner. 2003. Intelligent sensor fusion: A graphical model approach. In *Acoustics, Speech, and Signal Processing (ICASSP '03), IEEE International Conference on*, vol. 6, 733–36. Piscataway, NJ: IEEE.

28. Wymeersch, H., U. Ferner, and M. Win. 2008. Cooperative Bayesian self-tracking for wireless networks. *IEEE Commun. Letters* 12 (7): 505–7.

29. Mao, Y., F. Kschischang, B. Li, and S. Pasupathy. 2005. A factor graph approach to link loss monitoring in wireless sensor networks. *IEEE J. Selected Areas in Commun.* 23:820–29.

30. Wei, W., C. Wan-Dong, W. Bei-Zhan, W. Ya-Ping, and T. Guang-Li. 2008. The factor graph approach for inferring link loss in MANET. In *Internet Computing in Science and Engineering (ICICSE '08), International Conference on*, 166–72. Piscataway, NJ: IEEE.

31. Anker, T., D. Dolev, and B. Hod. 2008. Belief propagation in wireless sensor networks: A practical approach. In *Proceedings of the Third International Conference on Wireless Algorithms, Systems, and Applications*, 466–79. New York: ACM Press.

32. Chen, J., Y. Wang, and J. Chen. 2006. A novel broadcast scheduling strategy using factor graphs and the sum-product algorithm. *IEEE Trans. Wireless Commun.* 5 (6): 1241–49.

33. Sudderth, E., A. Ihler, W. Freeman, and A. Willsky. 2003. Nonparametric belief propagation. In *Computer Vision and Pattern Recognition, Proc. 2003 IEEE Computer Society Conf. on*, vol. 1, 605–12. Piscataway, NJ: IEEE.

34. Silverman, B. 1986. *Density estimation for statistics and data analysis*. Boca Raton, FL: Chapman and Hall/CRC.

35. Msechu, E., S. Roumeliotis, A. Ribeiro, and G. Giannakis. 2008. Decentralized quantized Kalman filtering with scalable communication cost. *IEEE Trans. Signal Process.* 56:3727–41.

36. Oka, A., and L. Lampe. 2009. Distributed scalable multi-target tracking with a wireless sensor network. In *Communications, 2009 (ICC '09), IEEE International Conference on*, 1–6. Piscataway, NJ: IEEE.

37. Li, R. X., and V. Jilkov. 2003. Survey of maneuvering target tracking: Part 1: Dynamic models. *IEEE Trans. Aerospace and Electronic Systems* 39 (4): 1333–64.

38. Liu, J., M. Chu, and J. Reich. 2007. Multitarget tracking in distributed sensor networks. *IEEE Signal Process. Mag.* 24 (3): 36–46.

39. Wei, C., L. Xin, and C. Mei. 2010. Cooperative distributed target tracking algorithm in mobile wireless sensor networks. In *Computer Science and Education (ICCSE), 5th International Conference on*, 120–23. Piscataway, NJ: IEEE.

40. Sandell, N., and R. Olfati-Saber. 2008. Distributed data association for multi-target tracking in sensor networks. In *Decision and Control (CDC 2008), 47th IEEE Conference on*, 1085–90. Piscataway, NJ: IEEE.

3 Implementation of Wireless Sensor Network Systems with PN-WSNA Approaches

Chung-Hsien Kuo and Ting-Shuo Chen

CONTENTS

3.1 INTRODUCTION

Wireless sensor networks (WSNs) [1–3] have been widely used to implement intelligent monitoring and control systems for a wide variety of applications. In general, WSN systems are composed of smart sensor nodes, which are deployed for autonomous sensing and decision making, such as factory automation [4], diagnosis [5], monitoring and control systems [6, 7], smart homes [8], etc. In practical terms, the most challenging aspect of WSN systems are the implementation and maintenance efforts needed for different domain-based software for the various application-oriented deployments of sensor nodes [9, 10].

WSN applications with different system characteristics, protocol designs, and application requirements make the task of domain engineers difficult in coping with particular hardware and software combination. Multidisciplinary experts have to collaborate to produce acceptable WSN solutions, e.g., in environmental health and air pollution studies. For example, the collaboration of environmental engineers and wireless network engineers is required to deal with domain-specific process models in order to produce appropriate sensor-node implementations.

Code maintenance is also an important task to be addressed by domain engineers, who usually rely on the help of information technology (IT) engineers, even for minor node software modifications. Furthermore, different sensor nodes may be burdened with different, distinct tasks that can be seen as different applications. It is challenging to maintain a wide variety of sensor applications that together work correctly and efficiently. To address these concerns, a systematic modeling approach combined with a model inference engine was proposed to help domain engineers implement and maintain their executable WSN systems without forcing them to handle the details of producing executable sensor-node code [10].

A model-based implementation also benefits the reconfiguration of WSN systems. For example, the reconfiguration of a WSN system is required when the control scenario in a remote sensor node is altered. Reconfigurations of remote sensor nodes are usually inconvenient because their software has to be recompiled and reloaded, typically by physically visiting and connecting to the sensor nodes. However, with a model-based implementation approach, the reconfiguration of a remote sensor node would be more flexible, potentially eliminating the need to recompile and reload the programs in the sensor node. Instead, it might be enough for a new, updated scenario model to be loaded to the data memory of the sensor node. The model inference engine, which acts as an executive program in a sensor node, would then interpret and execute the revised model scenario without stopping the execution of the sensor node.

In this chapter, a Petri-net-based wireless sensor-node architecture (PN-WSNA) is introduced to model the execution behaviors of a sensor node according to the tasks it needs to accomplish. The PN-WSNA is a high-level Petri net [11], and it is defined by inheriting properties of the ordinary Petri net (PN) [12, 13]; hence, the PN-WSNA–based implementation approach is capable of being used to evaluate the theoretical property that an implementation satisfies. Basically, the PN-WSNA model is composed of a number of events and conditions. Events are represented as transitions; conditions are represented as places. Arcs are used to describe pre- and post-conditions between places and transitions. Therefore, the conditions, events, and operational scenarios in a WSN system could be properly modeled using the PN-WSNA, where sensor events are generated in terms of the changes of sensor conditions.

A PN-WSNA model inference engine is realized and embedded in the sensor node as a virtual machine [14, 15], and it is capable of accessing and controlling the analog and digital I/O interfaces, as well as communicating with other senor nodes via a standard protocol, which in this case is ZigBee. Therefore, the domain engineers do not need to deal with detailed programming and I/O accessing natively on the sensor platform; instead, they just need to create the operational scenario models and to define the interfaces between the physical sensor and the sensor-node controller.

With the benefits of virtual machine and model-based implementation approaches, the reconfiguration of existing sensor nodes is accomplished by reloading the

PN-WSNA models, and no physical sensor-node visits are needed. More importantly, the PN-WSNA models can be evaluated for the properties of liveness, reachability, and boundedness before they are deployed [16]. Therefore, the PN-WSNA–based system is more reliable than conventional sensor nodes that are implemented ad hoc without any program validation. Meanwhile, the maintenance of models is more practical than the maintenance of platform-specific native code.

The rest of this chapter is organized as follows: Section 3.2 introduces the PN-WSNA; Section 3.3 illustrates the modeling of PN-WSNA and theoretical equivalencies; Section 3.4 demonstrates some PN-WSNA examples; Section 3.5 describes the user interface of the PN-WSNA integrated development environment (IDE); and finally, the conclusions are presented in Section 3.6.

3.2 PETRI-NET-BASED WIRELESS SENSOR NODE ARCHITECTURE (PN-WSNA)

Petri net (PN) is a modeling approach for discrete-event dynamic systems (DEDSs), and it was defined by Dr. Carl Adam Petri in 1962. The PN is capable of modeling events and conditions of systems in terms of causal relationships between events using mathematical and graphical representations. The ordinary PN structure is a four-tuple structure: $PN = (P, T, I, O)$, where P is a finite set of places; T is a finite set of transitions; I is the input function; and O is the output functions. Due to its ability to model concurrent and asynchronous system characteristics, the PN is capable of modeling the behaviors of elaborate monitor-and-control systems.

From the definition of PN, it is evident that the modeling efficiency of PN is problematic when the system scale is large or if the system is complex. Hence, various high-level PNs have been proposed to deal with systems with different characteristics, such as colored PNs [17, 18], timed PNs [19, 20], fuzzy PNs [21, 22], distributed object- and agent-oriented PNs [11, 23], etc. This chapter discusses the concepts of PN-WSNA. The proposed PN-WSNA is developed by inheriting the definitions of an ordinary PN. In order to deal with real-time sensor data acquisition, communication, and actuator control, additional interfaces and communication places are defined based upon an ordinary PN, with the intent of capturing the needs of practical WSN applications.

In addition to the interface and communication places, periodic sampling and execution tasks of the system are also defined using timed transitions. The sensor data is further categorized as high-enable and low-enable situations. Therefore, the PN-WSNA structure is an 11-tuple structure, $PN = (P_0, P_s, P_{ci}, P_{co}, P_a, T_0, T_t, T_H, T_L, I, O)$, where P_0 is a finite set of normal places; P_s is a finite set of sensor places; P_{ci} is a finite set of receiver places (communication incoming); P_{co} is a finite set of transmitter places (communication outgoing); P_a is a finite set of actuation places; T_0 is a finite set of immediate transitions; T_t is a finite set of timed transitions; T_H is a finite set of high-enable transitions; T_L is a finite set of low-enable transitions; I is the input function; and O is the output functions. The PN-WSNA graphical

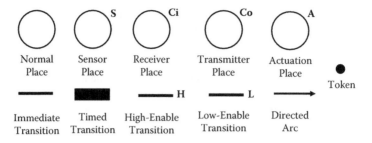

FIGURE 3.1 PN-WSNA graphical definitions.

definitions are shown in Figure 3.1. The PN-WSNA definitions are further elaborated as follows [10]:

1. *Place*: $P = \{p_1, p_2, p_3, ..., p_n\}$:$P$ is a finite set of places, $n \geq 1$. Places of the PN-WSNA are composed of normal places, sensor places, receiver places, transmitter places, and actuation places.

   a. *Normal places*: The definition of a normal place is the same as the place defined in the ordinary PN. Tokens in normal places may represent the status, condition, command, etc.

   b. *Sensor places*: The sensor place is defined to present the sensor status. One and only one token is initially assigned to each sensor place to enable sensor data collection. The sensor place may collect the sensor signals with different forms such as analog value, binary digits, or serial communication packets. The sensor data of p_i is denoted as $\eta(p_i)$. A sensor interface corresponds to an analog-digital-converter (ADC) channel, a generalized input-output (GIO) bit, or the universal asynchronous receiver/transmitter (UART) interface, and they are used depending on the support of the underlying hardware of the sensor nodes. A threshold value is defined for the sensor place to represent the analog sensor value into *high* or *low* status, and the threshold value of a p_i is denoted as $\varphi(p_i)$. The status of *high* and *low* is determined via the conflict output transitions (high-enable and low-enable transitions). Detailed descriptions of high-enable and low-enable transitions are described later.

   c. *Receiver and transmitter places*: A WSN system could consist of a large number of sensor nodes; as a consequence, communications among sensor nodes is also required to be represented. A receiver place and a transmitter place, used on different sensor nodes, can form a communication pair. A token in a transmitter place is immediately delivered to a corresponding receiver place. Hence, the transmitter place is a sink place [24], and the receiver place is a source place [24].

   d. *Actuation places*: An actuation place directly connects to a physical device, and it serves as an interface to drive peripheral devices and actuators. The actuation place plays a role similar to a transmitter place; however, the tokens in an actuation place are converted as actuation signals to control peripheral devices. Hence, an actuation place is also a sink place. Any token in an actuation place directly controls a peripheral device, and then the token is released. Possible actuation interfaces could be, for example, DAC0~DAC1, GIO1~GIO4, and UART.

2. *Transitions*: $T = \{t_1, t_2, t_3, ..., t_m\}$: T is a finite set of transitions, $m \geq 1$. Transitions may be used to define the events, time-based operations, and decisions. The execution (firing) of a transition may result in token migration in its input and output places. Transitions of the PN-WSNA are further classified as immediate transitions, timed transitions, and high- and low-enable transitions.

   a. *Immediate transitions*: The definition of an immediate transition is the same as the transition defined in the ordinary PN, and it can be used to model causal events and decisions. Tokens in the input places with directed arcs of an immediate transition would be immediately delivered to its output places when the immediate transition fires.

   b. *Timed transitions*: The definition of a timed transition is similar to the transition defined in the ordinary PN; however, a time that needs to elapse for firing is further defined. Tokens in the input places of a timed transition are not delivered to its output places immediately. Instead, a fired transition keeps the tokens until the predefined elapsed time has expired. Once the firing time is expired, tokens are released to its output places according to the output functions.

   c. *High-enable and low-enable transitions*: High-enable and low-enable transitions are especially defined for sensor places. High-enable and low-enable transitions are output transitions of a sensor place, and they must appear in a pair configuration. Conflicts of these transitions can happen. The firing of conflict high-enable and low-enable transitions depends on the sensor and threshold value defined in the sensor place. A high-enable transition is fired when the sensor data is greater than or equal to the threshold value defined in the input sensor place; a low-enable transition is fired when the sensor data is less than the threshold value defined in the sensor place.

3. In a PN-WSNA model, $P \cap T = \cdot$, and $P \cup T \neq ⁇$, where P is a finite set of places; T is a finite set of transitions.

4. *Token*: Tokens are quantitative representations of places, and they may represent the status (e.g., ready state of a sensor), conditions, and commands of systems.

5. *Marking and initial marking*: The marking is denoted as μ, which represents the token distributions in all places. μ is a $q \times 1$ column vector, and the jth element of μ indicates the number of tokens in place j, where q is a nonnegative integer, and it is equal to the number of places in a PN-WSNA model. The marking changes with respect to the firing of transitions if the PN-WSNA model is live. The initial marking (μ_0) is defined as the marking at system startup.

6. *Input and output functions, enabling, and firing*: Input and output functions are defined via directed and inhibited arcs. The input and output functions

are $I(p_i,t_j) \rightarrow N_{i,j}$ and $O(p_r,t_s) \rightarrow N_{r,s}$, respectively. $N_{i,j}$ and $N_{r,s}$ are nonnegative integers, and they are pre- and post-conditions of transitions.

7. *Enabling*: The enabling of a transition is further defined with the input functions of all directed arcs. A transition (t_j) is said to be enabled when it satisfies Equation (3.1).

$$\prod_{i=1}^{k} (I(p_i,t_j) - N_{i,j}) > 0, \qquad (3.1)$$

where $i = 1$ to k, and k is the number of input places of t_j; $p_i \in$ input places of t_j.

8. *Firing*: An enabled transition is not necessarily fired because of conflict situations. The conflict exists when the number of enabled transitions for a place is greater than 1. With a conflict situation, only one among all enabled transitions can be fired. Conflict transitions are resolved according to the following conditions:

   a. Random selections among enabled and conflict transitions are desired for immediate and timed transitions.

   b. To resolve the conflict situation of a pair of high-enable and low-enable transitions, the sensor value, $\eta(p_i)$, and threshold value, $\varphi(p_i)$, are evaluated for p_i. A high-enable transition can be fired if Equation (3.2) is satisfied.

$$\eta(p_i) \geq \varphi(p_i) \qquad (3.2)$$

A low-enable transition can be fired if Equation (3.3) is satisfied.

$$\eta(p_i) < \varphi(p_i) \qquad (3.3)$$

9. *PN-WSNA property evaluations*: The newly defined components of the PN-WSNA can be properly converted to PN components so that the PN-WSNA model property can be evaluated based on the same techniques as ordinary PN [24]. One can determine properties such as the incidence matrix, reachability tree, liveness, deadlock, boundedness, and safeness. The properties can be evaluated before the sensor-node application is deployed.

3.3 MODELING WITH PN-WSNA AND THEORETICAL EQUIVALENCES

In this section, several fundamental PN-WSNA models of sensor places, high/low-enable transitions, transmitter and receiver places, actuation places, and timed transitions are presented. Practically, PN-WSNA models are difficult for theoretical evaluation because of the extended definitions they include. Hence, model conversion approaches are proposed to convert the extended definitions of the PN-WSNA to ordinary PNs. Although the PN-WSNA model can be converted to ordinary PN models, complex PN models are also difficult to handle in terms of determining their

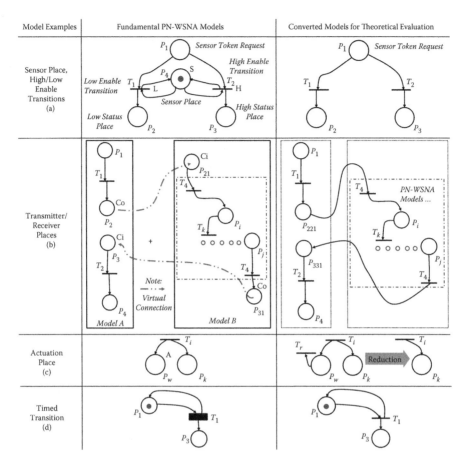

FIGURE 3.2   Elementary PN-WSNA models and equivalent PN models.

theoretical properties. Therefore, PN model-reduction approaches [13, 16] are used to reduce the PN model complexity as well as to reduce the efforts of theoretical evaluations without changing the model's properties. These fundamental PN-WSNA models and their equivalent theoretical PN models are summarized in Figure 3.2.

The first part introduces the elementary model of a sensor place and its PN conversion, as shown in Figure 3.2a. A token is initially assigned to a sensor place (P4) for the availability of sensor data collection. T1 and T2 are low-enable and high-enable transitions, respectively, and they are in conflict with each other. The firing conditions in Equations (3.2) and (3.3) are used to derive the sensor status. A normal place (P1) is used in this model to activate the execution of sensor data collection. Finally, two normal places (P2 and P3) are further used in this model to denote the result of "high" or "low" status.

From the definition of the sensor place, a sensor place is safe with a token. Therefore, the theoretical property of a sensor place could be the same as a place defined in the ordinary PN. Furthermore, according to the PN reduction rules [13, 16, 24], P4 and its inward and outward arcs can be removed to reduce the complexity of model evaluation. The high-enable and low-enable transitions theoretically behave the same role with the conflict transitions of an ordinary PN; hence, they can be directly converted to immediate transitions.

The second part is the elementary PN-WSNA model for communication, as shown in Figure 3.2b. The transmitter and receiver places are arranged in a pair configuration. Two PN-WSNA models (Model A and Model B) are connected in a cascade manner, and the token migrations between the two models are achieved in terms of transmitter and receiver places. In this example, P2 in Model A indicates a transmitter place that delivers tokens to a receiver place (P21) in Model B. The firing of Model B may finally generate a token in a transmitter place (P31). Then, the token in P31 is immediately delivered to a receiver place (P3) of Model A. It is noted that P_i, P_j, and T_k are intermediate places and transitions in Model B. Hence, P2 and P21 can be combined as P221; P3 and P31 can be also combined as P331. According to these conversions, two cascade PN-WSNA models can be properly converted as an ordinary PN model.

The third part describes the elementary PN-WSNA model for actuation places and their conversion, as shown in Figure 3.2c. In this model, P_w indicates an actuation place, and it directly corresponds to a physical actuation device. A token in P_w will be immediately converted to a corresponding control command to activate a connected device, and then the token is removed from the actuation place. Therefore, the actuation place is a sink place. In order to maintain the boundedness of the PN-WSNA, the theoretical evaluation is done via adding a sink transition to the output of P_w, and a converted ordinary PN model is formed. It is noted that the converted PN model can be further reduced to a simpler PN model by removing P_w and T_r.

Finally, the timed transition is usually used to represent the periodic sampling of the PN-WSNA–based sensor node. It is also helpful for representing the duty cycle of a node so that power consumptions and computational loads could be reduced. A typical sampling application is shown in Figure 3.2d. The timed transition behaves with the same enabling and firing rules as an ordinary PN transition except for the longer firing delay in timed transitions. Therefore, a timed transition can be theoretically evaluated in terms of approaching the elapsed time to zero. That is, the timed transition can be treated as an immediate transition for property evaluation purposes. Table 3.1 provides a summary of the PN-WSNA elementary component model definitions and their corresponding conversions.

3.4 EXAMPLES OF PN-WSNA MODELS

3.4.1 An Example Model

In this section, a simple 2-to-4 multiplexer PN-WSNA logic model is proposed for evaluating a combinational reasoning system. The 2-to-4 multiplexer is used to generate four decisions according to the binary status of two input signals. The input signals can be collected from two sensor devices. Hence, the sensor places model the sensors. The logic circuit of a 2-to-4 multiplexer is shown in the upper-left part

TABLE 3.1

PN-WSNA Conversion Summaries

Component	Approaches
P_s	Definition I: A sensor place with its inward and outward arcs can be removed for property evaluation
P_{ci}, P_{co}	Definition II: A pair of P_{ci} and P_{co} can be converted to a normal place for property evaluation
P_a	Definition III: An actuation place can be converted to a normal place and a sink transition for property evaluation
T_t	Definition IV: A timed transition can be converted to an immediate transition for property evaluation
T_H, T_L	Definition V: High-enable and low-enable transitions can be converted to immediate transitions for property evaluation

of Figure 3.3. It is noted that the symbol (*) indicates the Boolean inversion. In this example, S_A and S_B indicate the sensors A and B, respectively. The execution scenario of the 2-to-4 multiplexer is to make four combinational decisions with respect to the status of S_A and S_B. It is noted that S_A is a sensor used in a local sensor node (N_A), and S_B is a remote sensor used at a different sensor node (N_B). By referring to the elementary PN-WSNA models shown in Figure 3.2, the 2-to-4 multiplexer model can be constructed as shown in Figure 3.3. In this model, the PN-WSNA components of sensor places, high/low-enable transitions, transmitter and receiver places, actuation places, and timed transitions are all used.

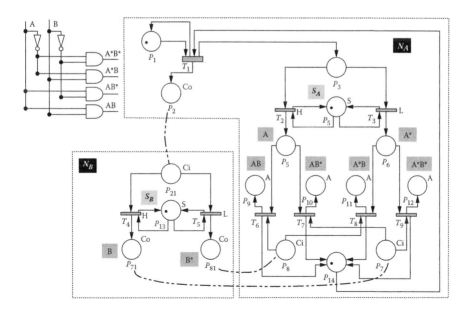

FIGURE 3.3   A 2-to-4 multiplexer PN-WSNA model with sensors A and B.

In Figure 3.3, two PN-WSNA models in sensor nodes N_A and N_B are created. The initial marking is to assign a token in P1, P4, P13, and P14, respectively. The first PN-WSNA model in N_A is used to model the 2-to-4 multiplexer with respect to a local sensor (S_A) and a remote sensor (S_B) in N_B. The second model is just used for sensor data collection of S_B. Moreover, the model in N_B is capable of receiving the request from N_A for data collection as well as to reply with the sensor status to N_A for the decision of a 2-to-4 multiplexer.

P1 indicates the availability of a 2-to-4 multiplexer PN-WSNA model in N_A. T1 is a timed transition for periodic sampling for both S_A and S_B. P2 and P3 are output places of T1. A transmitter place P2 is needed to deliver requests to a receiver place P21 so that S_B can be activated simultaneously. Two elementary sensor place models (as shown in Figure 3.2a) of P21→{T4,P13,T5}→{P71,P81} and P3→{T2,P4,T3}→{P5,P6} are used in both sensor nodes. Tokens in P4 and P13 represent the availability of S_A and S_B, respectively. P21 and P3 are used to activate the sensor data collection of S_A and S_B. T2 and T4 are high-enable transitions; T3 and T5 are low-enable transitions. The threshold values for high or low status are specified in S_A and S_B. If the sensor status is high, T2/T4 can be fired, and a token will be delivered to P5/P71. In terms of logic, P5 indicates A; P6 indicates A*; P6 indicates A; P71 indicates B; and P81 indicates B*. It is noted that the token in P71/P81 will be immediately transmitted to P7/P8.

Transitions of T6, T7, T8, and T9 are used to model the logic *AND* gates for the 2-to-4 multiplexer logic model according to the token appearing in P5, P6, P7, and P8. Notably, only two places among them have a token. Therefore, the decision can be taken by evaluating the combinations of A, A*, B, and B*. The results of AB, AB*, A*B, and A*B* are concluded in the actuation places of P9, P10, P11, and P12, respectively. These actuation places are further used to control predefined actuation devices, such as lamps. Meanwhile, the firing of T6, T7, T8, or T9 will release a token in P13 to indicate the finish of inference for a 2-to-4 multiplexer PN-WSNA model, and P13 and P1 will enable and fire T1 again to execute a new inference of this 2-to-4 multiplexer PN-WSNA model.

3.4.2 MODEL CONVERSIONS

The models shown in Figure 3.3 can be further evaluated in terms of PN model conversions and reductions. By referring to the PN-WSNA conversion approaches of Table 3.1, the PN-WSNA shown in Figure 3.3 can be properly converted as an ordinary PN model, as shown in Figure 3.4. The conversions are performed by eliminating the receiver places from a transmitter and receiver place pair, such as P2, P7, and P8. In addition, the actuation places were converted to normal places (from P9 to P12) with output transitions (from T10 to T13). In this manner, the converted model in Figure 3.4 is a timed PN model. The converted model can be further reduced to a simpler model without changing its properties, as shown in Figure 3.5. Hence, the PN model in Figure 3.5 is theoretically equivalent to the PN-WSNA model indicated in Figure 3.3. It is noted that in order to evaluate the theoretical properties, the numbering of transitions and places are reorganized.

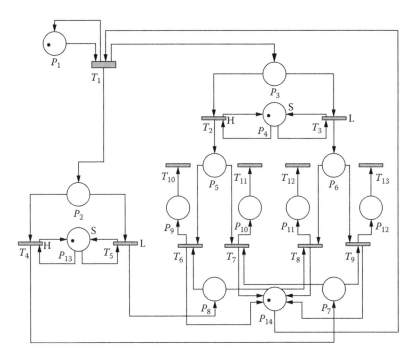

FIGURE 3.4 Converting the PN-WSNA model of Figure 3.3 to a timed PN model.

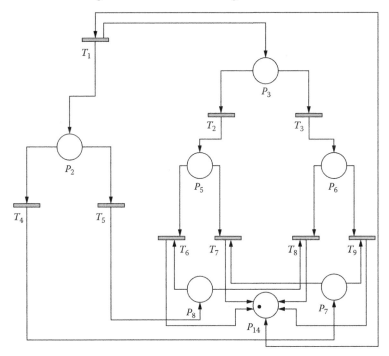

FIGURE 3.5 Theoretical equivalent conversion of Figure 3.4.

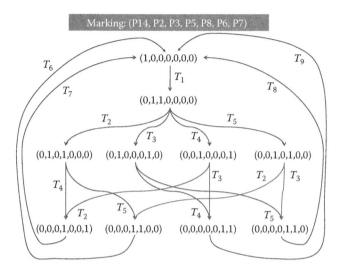

FIGURE 3.6 Reachability graph of Figure 3.5.

3.4.3 PROPERTY EVALUATIONS

Based on the theoretical equivalence model, the reachability graph can be evaluated, as shown in Figure 3.6. The marking is arranged as the sequence of (P14, P2, P3, P5, P8, P6, P7). The theoretical equivalence model exhibits 10 states (markings). Each marking is bounded and safe. In addition, this model is also live. Moreover, the incidence matrix is also evaluated based on the operation of $D^+ - D^-$, where D^+ and D^- are as indicated in Equation (3.4). Based on Equation (3.4), a reachable state can be obtained from the incidence matrix, current marking, and firing transitions, as noted in the work of Murata [24]. As a consequence, the example 2-to-4 multiplexer PN-WSNA model, which involves two sensor nodes that can be evaluated together, is live, safe, and deadlock free.

$$
D^- = \begin{bmatrix}
1 & 0 & 0 & 0 & 0 & 0 & 0 \\
0 & 0 & 1 & 0 & 0 & 0 & 0 \\
0 & 0 & 1 & 0 & 0 & 0 & 0 \\
0 & 1 & 0 & 0 & 0 & 0 & 0 \\
0 & 1 & 0 & 0 & 0 & 0 & 0 \\
0 & 0 & 0 & 1 & 1 & 0 & 0 \\
0 & 0 & 0 & 1 & 0 & 0 & 1 \\
0 & 0 & 0 & 0 & 1 & 1 & 0 \\
0 & 0 & 0 & 0 & 0 & 1 & 1
\end{bmatrix}; \quad
D^+ = \begin{bmatrix}
0 & 1 & 1 & 0 & 0 & 0 & 0 \\
0 & 0 & 0 & 1 & 0 & 0 & 0 \\
0 & 0 & 0 & 0 & 0 & 1 & 0 \\
0 & 0 & 0 & 0 & 0 & 0 & 1 \\
0 & 0 & 0 & 0 & 1 & 0 & 0 \\
1 & 0 & 0 & 0 & 0 & 0 & 0 \\
1 & 0 & 0 & 0 & 0 & 0 & 0 \\
1 & 0 & 0 & 0 & 0 & 0 & 0 \\
1 & 0 & 0 & 0 & 0 & 0 & 0
\end{bmatrix}
\qquad (3.4)
$$

3.5 PN-WSNA INTEGRATED DEVELOPMENT ENVIRONMENT

In addition to the definition of PN-WSNA, this chapter also introduces the PN-WSNA integrated development environment (IDE). The PN-WSNA IDE is developed for constructing a PN-WSNA–based autonomous sensor system in terms of a model-based implementation approach. The PN-WSNA system is composed of a PN-WSNA management server and a number of autonomous sensor nodes. A detailed implementation of PN-WSAN–based autonomous sensor nodes can be found in the literature [10]. The PN-WSNA management server is an IDE, and it provides a graphical user interface for fast model construction and reconfiguration without physical sensor-node visits. Figure 3.7 shows the PN-WSNA IDE, and the toolbar icons are used for model constructions, editing, revisions, manipulations, run-time simulations, model drawing auxiliaries, as well as for model loading and reconfiguration of remote sensor nodes. Their functions are summarized in Table 3.2.

In order to perform the PN-WSNA model's inferences for both run-time simulations and sensor-node kernel execution, the model and run-time attributes of the places, transitions, and arcs are further defined such that the model attributes are static, which is desired for the sake of the model structure; the run-time attributes are dynamic, and they are used for the model inference. These attributes can be assigned by using the property dialogs, as shown in Figure 3.8. Finally, a summary of these attributes is shown in Table 3.3. It is noted that the attached message is used for the reporting system within the node ID, place ID, token number, and time stamp. In addition, the sensor data can also be combined with the attached message for sensor places if the *attach sensor data* item is selected.

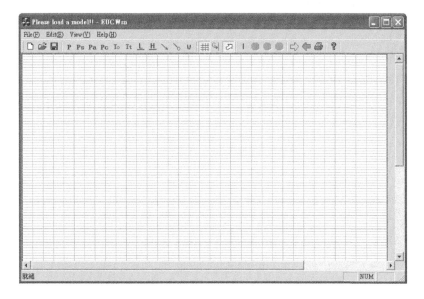

FIGURE 3.7 PN-WSNA IDE workspace.

TABLE 3.2
PN-WSNA IDE Toolbar Descriptions

Icon	Descriptions
P	Add a new normal place
Ps	Add a new sensor place
Pa	Add a new actuation place
Pc	Add a new transmitter or receiver place (depending on "Output Mode" check box of a place setup dialog)
Ic	Add a new immediate transition
It	Add a new timed transition
L	Add a new low-enable transition
H	Add a new high-enable transition
↘	Add a new directed arc between a pair of place and transition
↘○	Add a new inhibited arc between a pair of place and transition
u	Add a notation in the drawing area
⊞	Show grids in the drawing area
⊡	Fit all PN-WSNA components in the drawing area to grids
⬚	Select a PN-WSNA component for editing, changing positions, deleting, checking setup
I	Initialize the simulation conditions (i.e., loading initial markings)
●	Simulate the model for one step (step-by-step firing)
G	Simulate the model continuously
●	Stop the simulation
↻	Reconfigure a sensor node
⬅	Retrieve a model from a sensor node

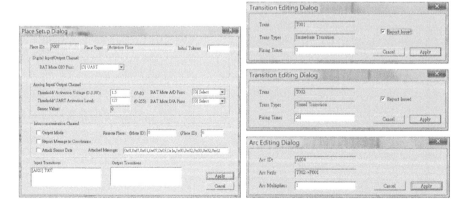

FIGURE 3.8 PN-WSNA place, transition, and arc setup dialogs.

TABLE 3.3
PN-WSNA Static and Dynamic Attributes

Symbol	Static and Dynamic Attributes
P	*Static:* place ID, place type, inward arcs, outward arcs, report issued, attached message, and sensor data that are attached to reports
	Dynamic: token number
P_s	*Static:* place ID, place type, inward arcs, outward arcs, report issued, attached message, sensor data that are attached to reports, interface type, interface pin number, ADC threshold value, and UART threshold value
	Dynamic: token number and sensor value
P_{ci}	*Static:* place ID, place type, outward arcs, report issued, attached message, and sensor data that are attached to reports
	Dynamic: token number
P_{co}	*Static:* place ID, place type, inward arcs, report issued, attached message, sensor data that are attached to reports, remote mote ID, and remote place ID
	Dynamic: token number
P_a	*Static:* place ID, place type, inward arcs, report issued, attached message, sensor data that are attached to reports, interface type, interface pin number, DAC control value, and UART packets
	Dynamic: token number
T_0, T_L, T_H	*Static:* transition ID, transition type, inward arcs, outward arcs, and report issued
	Dynamic: status
T_t	*Static:* transition ID, transition type, firing time, inward arcs, outward arcs, and report issued
	Dynamic: status, remaining firing time
I, O	*Static:* arc ID, arc multiplier, input function, output function
	Dynamic: N/A

3.6 CONCLUSIONS

In this chapter, a PN-WSNA model-based implementation approach is introduced for implementing wireless senor network systems. The advantages of model-based approaches include fast system implementation, increased maintainability, and reliable system operations without the need to consider the details of low-level platform-specific native code implementation. In addition, the PN-WSNA also provides the ability to explore theoretical properties of the model, e.g., for boundedness and liveness investigation, as well as for run-time simulations of the WSN execution scenarios before the sensor models are deployed. With the PN-WSNA–based WSN system implementation approaches, the WSN engineers can focus on the execution scenario models instead of programming complicated platform-specific applications. An example 2-to-4 multiplexer PN-WSNA logic model built over two communicating sensor nodes was introduced, and its properties were evaluated. In the future, the PN-WSNA approach is planned to be used to implement production systems such as autonomous multirobot decision systems, smart home systems, etc.

ACKNOWLEDGMENTS

This work was supported by the National Science Council, Taiwan, R.O.C., under Grant NSC 100-2221-E-011 -022.

REFERENCES

1. Romer, K., and F. Mattern. 2004. The design space of wireless sensor networks. *IEEE Wireless Communications* 11 (6): 54–61.
2. Akyildiz, I. F., T. Melodia, and K. R. Chowdhury. 2008. Wireless multimedia sensor networks: Applications and testbeds. *Proceedings of the IEEE* 96 (10): 1588–1605.
3. Dalola, S., V. Ferrari, M. Guizzetti, D. Marioli, E. Sardini, M. Serpelloni, and A. Taroni. 2009. Autonomous sensor system with power harvesting for telemetric temperature measurements of pipes. *IEEE Trans. Instrumentation and Measurement* 58 (5): 1471–78.
4. Zhuang, L. Q., W. Liu, J. B. Zhang, D. H. Zhang, and I. Kamajaya. 2008. Distributed asset tracking using wireless sensor network. In *IEEE International Conference on Emerging Technologies and Factory Automation*, 1165–68. Piscataway, NJ: IEEE Press.
5. Ballal, P., A. Ramani, M. Middleton, C. McMurrough, A. Athamneh, W. Lee, C. Kwan, and F. Lewis. 2009. Mechanical fault diagnosis using wireless sensor networks and a two-stage neural network classifier. In *IEEE Aerospace Conference Proceedings*, 1–10. Piscataway, NJ: IEEE Press.
6. Sridhar, P., A. M. Madni, and M. Jamshidi. 2007. Hierarchical aggregation and intelligent monitoring and control in fault-tolerant wireless sensor networks. *IEEE Systems Journal* 1 (1): 38–54.
7. Tsow, F., E. Forzani, A. Rai, R. Wang, R. Tsui, S. Mastroianni, C. Knobbe, A. J. Gandolfi, and N. J. Tao. 2009. A wearable and wireless sensor system for real-time monitoring of toxic environmental volatile organic compounds. *IEEE Sensors Journal* 9 (12): 1734–40.
8. Suh, C., and Y.-B. Ko. 2008. Design and implementation of intelligent home control systems based on active sensor networks. *IEEE Transactions on Consumer Electronics* 54 (3): 1177–84.
9. Kuo, C. H., and J. W. Siao. 2009. Petri net based reconfigurable wireless sensor networks for intelligent monitoring systems. In *International Conference on Computational Science and Engineering*, Vol. 2, 897–902. Piscataway, NJ: IEEE Press.
10. Kuo, C. H., and T. S. Chen. 2011. PN-WSNA: An approach for reconfigurable cognitive sensor network implementations. *IEEE Sensors Journal* 11 (2): 319–34.
11. Kuo, C. H., C. H. Wang, and K. W. Huang. 2003. Behavior modeling and control of 300 mm fab intrabays using distributed agent oriented Petri net. *IEEE Trans. Systems, Man, and Cybernetics, Part A* 33 (5): 641–48.
12. Li, Z. W., M. C. Zhou, and M. D. Jeng. 2008. A maximally permissive deadlock prevention policy for FMS based on Petri net siphon control and the theory of regions. *IEEE Trans. Automation Science and Engineering* 5 (1): 182–88.
13. Koh, I., and F. DiCesare. 1990. Transformation methods for generalized Petri nets and their applications to flexible manufacturing systems. In *Proceedings of Rensselaer's Second International Conference on Computer Integrated Manufacturing*, 364–71. Piscataway, NJ: IEEE Press.
14. Avvenuti, M., P. Corsini, P. Masci, and A. Vecchio. 2007. An application adaptation layer for wireless sensor networks. *Pervasive and Mobile Computing* 3 (4): 413–38.
15. Levis, P., and D. Culler. 2002. Mate: A tiny virtual machine for sensor networks. In *Proceedings of International Conference on Architectural Support for Programming Languages and Operating Systems*, 85–95. New York: ACM Press.

16. Shatz, S. M., S. Tu, T. Murata, and S. Duri. 1996. An application of Petri net reduction for Ada tasking deadlock analysis. *IEEE Trans. Parallel and Distributed Systems* 7 (12): 1307–22.
17. Wu, S. M., and S. J. Lee. 1997. Enhanced high-level Petri nets with multiple colors for knowledge verification/validation of rule-based expert systems. *IEEE Trans. Systems, Man, and Cybernetics, Part B: Cybernetics* 27 (5): 760–73.
18. Chiola, G., C. Dutheillet, G. Franceschinis, and S. Haddad. 1993. Stochastic well-formed colored nets and symmetric modeling applications. *IEEE Trans. Computers* 42 (11): 1343–60.
19. Holliday, M. A., and M. K. Vernon. 1987. A generalized timed Petri net model for performance analysis. *IEEE Trans. Software Engineering* 13 (12): 1297–1310.
20. Hu, H., M. C. Zhou, and Z. Li. 2010. Low-cost and high-performance supervision in ratio-enforced automated manufacturing systems using timed Petri nets. *IEEE Trans. Automation Science and Engineering* 7 (4): 933–44.
21. Lee, J., K. F. R. Liu, and W. Chiang. 1999. A fuzzy Petri net-based expert system and its application to damage assessment of bridges. *IEEE Trans. Systems, Man, and Cybernetics, Part B: Cybernetics* 29 (3): 350–70.
22. Shen, V. R. L. 2006. Knowledge representation using high-level fuzzy Petri nets. *IEEE Trans. Systems, Man, and Cybernetics, Part A: Systems and Humans* 36 (6): 1220–27.
23. Kuo, C. H., and H. P. Huang. 2003. Distributed performance evaluation of a controlled IC Fab. *IEEE Trans. Robotics and Automation* 19 (6): 1027–33.
24. Murata, T. 1989. Petri nets: Properties, analysis and applications. *Proceedings of the IEEE* 77 (4): 541–80.

4 Real-Time Search in the Sensor Internet

Richard Mietz and Kay Römer

CONTENTS

4.1   INTRODUCTION

The term *Internet of Things* was coined in the late '90s by Kevin Ashton. Part of this vision is the integration of small, networked, inexpensive embedded computing devices throughout our living environment to ease our everyday lives. Many of the envisioned devices, such as smart phones and wireless sensor networks, will have sensory capabilities to perceive the state of their surroundings. These sensors create a foundation for new applications and services to interact with the real world. One key aspect is to search for and find an adequate subset of sensors that provide the user with the desired functionality and data. This chapter discusses the challenge of searching in sensor networks by presenting the general concepts of Web search along with the drawbacks when adopted for sensor networks as well as new approaches to overcome the resource limitations of small embedded devices.

Two decades ago the Web evolved, and with more and more Web resources available, the need for a search service to find suitable resources on a specific topic also arose. At that time, Web content was rather static, meaning that resources were updated at long intervals or never even changed. Broder [1] gives a first simple classification of Web search, dividing it into navigational, informational, and transactional search. With a navigational query, the user tries to find a specific website, while the intent of the second type, informational queries, is to find one or more resources with desired information. The last one, transactional query, includes searches with the goal to find pages where additional interaction is required. Examples of such searches are file downloads and shopping pages. This initial static Web is often referred to as the Web 1.0.

The term *Web 2.0* was coined around 2004 and describes the next step in Web history, with a paradigm shift in Web usage where the border between producers and consumers blurs: Everybody can easily publish new content such as text messages, pictures, and videos using platforms like Twitter, Flickr, and YouTube. Additionally, websites with automatically generated content, such as stock and weather portals, are updated more frequently. Hence, users no longer only search for static content, but also look for dynamic content like stock exchange prices, weather, and recent tweets on a specific topic. Moreover, search behavior changes toward location-centric search [2], i.e., users try to find resources like doctors and restaurants in their proximity. This is also reflected in a more detailed classification of Web search goals [3]. So, while search in the Web 1.0 can be described as *static search*, Web 2.0 search is more diverse and thus can be characterized as *real-time search*, *dynamic search*, and *location-centric search*.

Certainly, Web 2.0 is not the end of the Web development. Web 3.0 is often used as a synonym for the more meaningful term *Semantic Web*, where semantic descriptions allow machines to understand the meaning and relation of resources and to infer new knowledge and relations from known facts. Another vision is the Web of Things/Sensor Web or, more generally, the Internet of Things, where everyday objects will be equipped with sensor devices and connected to the Internet, thus enabling them to publish their current state on the Web. Already today, many electronic devices such as smart phones and alarm clocks are augmented with sensors for measuring temperature, humidity, acceleration, or light intensity. In addition, more and more sensor networks performing environmental, structural health, or traffic monitoring are deployed and connected to the Internet. Thus, the state of an increasing fraction of the real world will be accessible online on the Web, and users are becoming more and more interested in combining location-centric search with search for real-world entities with a certain state in real time. Examples are the search for *an empty meeting room in the university, a cinema showing a certain movie with available seats*, or even queries referring to multiple real-world entities such as for *a free parking lot in vicinity of an Italian restaurant with free tables and low noise level within five kilometers of the current position*. Such a Web search can be characterized as a *real-time, real-world, location-centric search*.

The concepts to search and find static information with keyword-based systems are well understood by industry and users and is used hundreds of millions times every day. Taking into account the possibilities of real-world search, these numbers will multiply. However, a new generation of search engines is needed that tackles the underlying challenges, which range from sensor integration and data gathering, to data processing, to more natural and user-friendly query languages.

In the next section, we give an overview of the techniques used by Web search engines today. The drawbacks of these techniques when used in combination with sensor networks, and possible approaches to overcome these problems, are discussed in Section 4.3. Finally, in Section 4.4 we present an architecture for a semantic sensor search engine as a first step toward a next-generation search engine.

4.2   SEARCH IN THE "TRADITIONAL" INTERNET

The concepts used by search engines today in the traditional Internet are tailored to the plentiful resources available from the systems providing the Web content, i.e., powerful and line-powered computers with stable, reliable, high-bandwidth Internet connections. Thus, optimization of energy consumption and bandwidth efficiency is not a key aspect of search engine providers when interacting with Web servers.

4.2.1   GENERAL CONCEPTS

A search engine has to realize three main functions: gathering Web content, processing this data to extract and store relevant information for fast lookup, and answering user queries to provide the most relevant results.

4.2.1.1   Gathering Web Content

The crawler of a search engine is responsible for gathering Web resources, as depicted in Figure 4.1. The Web can be considered as a directed graph with websites as nodes and the embedded hyperlinks pointing to other websites as the directed edges. Crawling is most often a breadth-first search on this Web graph, but other approaches such as backlink-count, where websites with a high number of links pointing to them are visited first, are also used.

The process begins with the *seed*, a list of URLs to visit first. While visiting, downloading, and storing the websites and often also their embedded elements

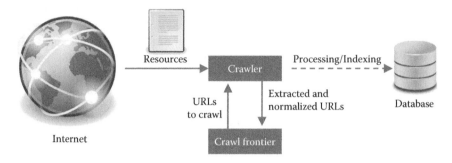

FIGURE 4.1 Workflow of a crawler.

such as images and videos, embedded URLs from the *href* and *src* HTML tags are extracted, normalized (e.g., conversing lowercase, deleting anchors, and sorting of URL parameters), and afterward added to the visiting list, called *crawl frontier*. The URL to visit next is selected from the crawl frontier by using a *selection policy* like the breadth-first search or the backlink count. Two other mechanisms affect the decision to visit a site or not. First, the crawler can look for a file named *robots.txt* in the root directory of the Web server, including rules indicating which files and directories should not be visited. The counterpart to the *robots.txt* is the *Sitemaps protocol** invented by Google. This XML file contains a list of websites along with optional parameters for the last modification time, the change frequency, and a priority relative to other URLs of that website. It can be uploaded to an arbitrary directory on the server and is referenced by a nonstandard field in *robots.txt*. The Sitemaps protocol is useful to point the crawler to part of the *Deep Web/Invisible Web*, e.g., websites not reachable by hyperlinks and dynamic Web content revealed by submitting a form.

Because an increasing number of websites are frequently updated and new pages are added often, the crawling process has to be repeated at faster intervals to ensure coverage and the freshness of gathered data. The interval to crawl a website for the next time is handled by the *revisit policy*. With a *uniform* policy, the revisit interval is the same for every website. Due to the enormous variability of the update frequency of websites, a uniform interval would have an adverse effect, either by unnecessary crawling of nonupdated pages or by missing changes of frequently updated websites. Hence, many search engines exploit a *proportional policy*, where the visiting interval for a page is proportional to its change frequency, i.e., the higher the update rate is, the more often the site is crawled. For Web pages with an extremely high update rate and thus a short revisiting interval, crawlers would consume a lot of resources by means of bandwidth and server load, leading to a bad user experience when using such websites. Hence, search engines employ a *politeness policy* to balance the revisiting interval and resource consumption of Web servers. While this policy influences the external resources such as bandwidth and Web server load, there is sometimes a fourth policy, the *parallelization policy*, which affects the internal resources such

---

* http://www.sitemaps.org/

as memory and processor. This policy determines how many and which pages are loaded simultaneously when using a parallel crawler. Hence, the policy also has to take care of URLs discovered by different processes such that a website is not loaded multiple times.

4.2.1.2 Indexing Web Content

Indexing is the process of extracting and storing information from websites such that queries can be efficiently answered. If the update frequency of a Web resource is lower than the query frequency, then the overhead of building an index is low compared to the resources needed when scanning every document at query time to identify matching websites.

The indexing process starts with extracting relevant information from the crawled websites. Therefore, first, the language of the website is determined. This information is needed to complete the following steps of tokenization, stop-word removal, and stemming. While tokenization, the process of splitting text into several tokens, i.e., words in this case, is easy for languages such as English, it is much harder, for example, for Chinese. Afterward, all words occurring with high frequency in the detected language, so-called stop words, are removed. Due to their high frequency, the probability of finding them in a document is rather high. Thus, using them in the search process would not improve results but, rather, tremendously increases storage needs. Finally, for every word, the stem is derived and used in future steps. As a consequence, different inflections of a word as part of a document or query can match each other. Besides word extraction, other information like formatting of elements and meta information such as title and publisher are extracted, and statistics such as word frequency are computed.

A forward index stores for every document the list of extracted words. So, the data structure is a list of mappings between a website and a word, grouped and sorted by the website. If the forward index is rearranged in the sense that it is sorted and grouped by words, a record-level inverted index is formed (see Figure 4.2). Typically, a word-level inverted index, i.e., a record-level inverted index with the position for each word in the document, is formed because it allows easier search for phrases and words with a certain proximity in the document. As billions of websites exist, the size of an inverted index can grow up to thousands of terabytes, as is the case for Google. Hence, compression is used to lessen storage at the cost of greater power consumption associated with the higher processor utilization for compression and

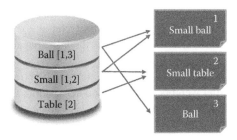

FIGURE 4.2 Inverted index.

decompression. The performance of update and delete operations of an inverted index is dependent on the underlying data structure. Usually, a tree data structure, such as a B-Tree, is used, resulting in O($log\ n$) runtime. With millions of entries in an inverted index, these operations can be rather expensive with respect to time and disc accesses.

4.2.1.3 Understanding User Intent

Due to the ambiguity of queries formulated in natural languages, search engines resort to keyword-based queries. However, the use of keywords reduces the expressiveness of queries, thus hampering the understanding of the user intent by the search engine. Hence, search engines provide additional mechanisms to support the user, such as searching for phrases. Several keywords and phrases can be combined with the help of Boolean expressions, allowing to find all given terms using an *AND* operator, at least one keyword or phrase with the help of an *OR* operator, or, with the help of a *NOT* operator, to specify terms that should not be contained in the result documents. Furthermore, the scope of resources to find can be limited to specific file types, languages, or the time since last modification. These options can be set via menus or directly as part of the textual query by using special terms in the form of *scope:value*, such as *filetype:pdf* for finding only documents in pdf format. Autocompletion of query terms while the user is typing is another support mechanism to help the user write meaningful queries. This technique is realized by asynchronously sending typed letters to the search engine backend, receiving keyword suggestions, and presenting these to the user.

Once a query is submitted, it is parsed, and extracted keywords are stemmed and finally used to find matching documents in the inverted index. In the case of several keywords, matching documents for each keyword are retrieved, and in a further step they are joined such that the only documents remaining contain all the query terms.

4.2.1.4 Presenting Results

The last step in query answering is the presentation of found results to the user. Usually the result set consists of thousands to millions of websites, and typically a user only browses the first dozens of results to find relevant websites for the initial query. Thus, an ordering that provides relevant results to the user is a key challenge when displaying results. Hence, each search engine uses its own, mostly secret, ranking algorithm with a different selection and weighting of factors. However, some often-used ranking factors are known from patents, search engine news, and analysis. Tables 4.1, 4.2, and 4.3 show some ranking factors related to, respectively, keywords of the query, links between sites, and other metrics at the page as well as the domain levels. Because website providers and spammers also know about the relevance of these factors, they try to prepare websites that achieve high ranking results. Search engines react by adjusting weights and including new factors from time to time.

Considering several factors when ranking websites consumes additional resources in terms of storage access and processing power. Hence, search engines cache results for frequently appearing queries, such as hot topics or famous persons, to minimize

TABLE 4.1
Keyword-Based Ranking Factors

Page Level	Domain Level
Total # of keyword occurrences	Keyword is domain name
Keyword in title	Keyword is part of domain name
Keyword in heading	Keyword is first word of domain name
Keyword in metadata	Keyword is subdomain name
Inflection of keyword	Keyword is part of subdomain name
Proximity of keywords	

Source: SEOmoz, http://www.seomoz.org/article/search-ranking-factors [4].

TABLE 4.2
Link-Based Ranking Factors

Page Level	Domain Level
Total # of links to website	# of unique domains linking to this domain
# of IPs linking to website	# of unique URLs linking to this domain
# of external links	Topical relevance of the domains linking to this domain
% of external links	"Distance" of the domain from a trusted seed in terms of link hops
# of internal links to website	

Source: SEOmoz, http://www.seomoz.org/article/search-ranking-factors [4].

TABLE 4.3
Further Ranking Factors

Page Level	Domain Level
Length of website	Length of domain name
Uniqueness of websites' content	Uniqueness of the domains' content
Age of website	Time since domain was registered
Freshness of website	Time since domain registration details have changed
Loading speed	Loading speed across the domain
Use of advertising	# of links to error pages from the domain
Validation errors	# of error pages under that domain

Source: SEOmoz, http://www.seomoz.org/article/search-ranking-factors [4].

lookup time and access to the inverted index. Cached results are updated regularly to reflect changes in the Web.

Finally, the ranked results are shown as a result page. Typically the title, the URL, and a text snippet emphasizing the keywords are presented for each result.

4.2.2 Systems

There are hundreds of search engines on the Web. We discuss different types of search engines and give examples for each category along with some distinctive features.

4.2.2.1 General Search Engines

Most of the search engines on the Web are general search engines aiming to search the complete Web. Example systems are the well-known search engines Google,[*] Microsoft's Bing,[†] Baidu[‡] (most popular search engine in China), and Yandex[§] (most popular search engine in Russia). While these search engines generally work as described in Section 4.2.1, each has its own ranking algorithm.

The most known ranking algorithm is PageRank, developed by the founders of Google in 1998 [5]. The algorithm calculates a numerical weight, the PageRank, for each website, indicating the relevance compared to the other pages. The idea behind PageRank is that a hyperlink to a website counts as a supporting vote for that website. A website receives a high PageRank if it is referenced by other websites with a high PageRank. Hence, the algorithm is recursive.

Initially, it was not hard to manipulate the results and thus gain good ranking results. Consequently, several improvements and other factors have been developed and are used today. The Block-Level-PageRank divides a website into more- and less-important areas. Links are weighted according to which of these areas they belong. Another improvement is TrustRank. In addition to the PageRank computation, the distance in link hops from a trusted seed of URLs is considered. Thus, spam sites are usually ranked lower because the distance from a trustful Web site is rather high.

Nowadays, Google incorporates several metrics from its other projects to measure the relevance of websites for users. For example, statistics of user behavior (e.g., how long a user stays on a website, which links a user follows, and which country the user comes from) is collected from Google's Chrome Browser, their advertising platform AdSense, and their website statistics tool Analytics.

Yandex developed a technology called Spectrum [6] to better understand ambiguous queries of users. By analyzing the user behavior (e.g., which links of the result are clicked) for identical queries in the past, the current intent is predicted and the ranking is adjusted. For example, if the majority of users is looking for information about a movie when searching for *Matrix*, search results about the film will be ranked higher than results about the mathematical object.

---

[*] http://www.google.com
[†] http://www.bing.com
[‡] http://www.baidu.com
[§] http://www.yandex.com

Another approach of Yandex is MatrixNet [7], a learning ranking algorithm. MatrixNet computes ranking formulas from thousands of factors while at the same time being resistant to overfitting. A characteristic feature is the possibility of constructing adjusted ranking formulas for a class of queries, e.g., for queries related to movies.

4.2.2.2 Specialized Search Engines

Besides general search engines covering resources of the whole Web, there are several specialized search engines. Because of the constricted field of search and specific type of resources, crawling policies and indexing processes of these search engines are optimized for the supported domain. Although we present different types of search engines along with example systems, not every system can be unambiguously assigned to a single category.

4.2.2.2.1 Topical Search Engines

Several search engines specialize their capabilities on particular entities such as images, videos, music, or papers; others concentrate on a specific topic like cooking or sports. Google, for example, runs several specialized search engines, e.g., for locations,[*] scientific work,[†] images, videos, and news.[‡]

Multimedia search engines use crawling techniques similar to those used by general search engines, but instead of concentrating on text content, they focus more on the embedded multimedia files. However, the selection policy of image crawlers controls which extracted images to keep for the following indexing process. Images that are too small or completely transparent are often discarded because they are usually used for design purposes of websites. Topical search engines use focused crawlers with a special selection policy that restricts the analysis to websites relevant for a specific topic. Some follow only URLs where the link description tag contains topic-matching keywords or where the website itself is about the topic, while others download all websites, analyze them, and discard irrelevant websites.

To find matching images or videos for keywords of a query, the search engine needs to guess the content of these files in the indexing process. The simplest method is to not examine the multimedia files themselves, but the website in which they are embedded. From this, text, especially near the embedded object, title, describing HTML tags, path, and filename are analyzed. However, this often yields insufficient information. Hence, many search engines investigate the file itself. Many multimedia files carry embedded metadata descriptions, such as the ID3-Tag of MP3 music files or EXIF data of JPEG and TIFF images, holding information about the creator, time of creation, location of creation, and title. If present, indexing this data field gives more detailed information about the file. Some search engines such as Google and Bing have already started to analyze the actual audio or image file. They can by now detect faces to allow search for images with persons in them. With more sophisticated algorithms and a bigger database, these techniques will also likely be used to classify other objects automatically in the future.

---

[*] http://maps.google.com
[†] http://scholar.google.com
[‡] http://news.google.com

When it comes to ranking, topical search engines use ranking factors similar to those used by general search engines, while multimedia search engines also consider factors such as length and bit rate of music and video files and resolution and color depth of images.

4.2.2.2.2 Semantic Search Engines

Semantic technologies are used in two different ways in search engines. Hakia[*] processes the query in a semantic fashion to better understand the intent of the user by using not only the given keywords to search in its data set, but also morphological variations, synonyms, generalizations, and similar concepts described by the keywords [8]. Additionally, it handles queries in natural language such as questions. In contrast, Freebase[†] uses semantics in its data set, i.e., it maintains a database of structured data and facts that are used to answer queries. New facts are either retrieved by crawling the Web for semantic data or by manual insertion by users. Of course, there are also search engines such as TrueKnowledge[‡] that combine both forms of semantics in their search.

Semantic data is described with the concepts of the Resource Description Framework (RDF) and Ontologies. RDF is a family of standards to describe facts in the form of Subject-Predicate-Object triples. Several triples form a graph with subjects and objects as nodes and predicates as directed edges, thus modeling relationships between different subjects and objects. Subjects and predicates are Resources, each encoded as a Uniform Resource Identifier (URI) to guarantee the uniqueness, while an object can be a Resource or a Literal, i.e., a string, date, or number. Ontologies allow to align different URIs describing the same concept or entity, i.e., with the help of an ontology, a computer can reason that the URIs are equal and thus exchangeable. With these descriptions and other inference rules, new facts can be derived from known facts. If a triple states that a *car* is a subtype of a *vehicle* and another triple states that a *cabrio* is a subtype of a *car*, the new fact that *cabrio* is a *vehicle* can be inferred.

As most Web content lacks semantic annotation, crawlers of semantic search engines, such as DBPedia,[§] a semantic encyclopedia, concentrate on sources of structured data. Instead of keyword extraction and maintenance of an inverted index, gathered triples are usually stored in a specialized database, called triplestore. SPARQL, a query language for the graph formed by the RDF data, is used to retrieve information from the triplestore.

Ranking is similar to general search engines for semantic search engines, which try to understand the user's intent. However, when the data set consists of structured semantic data, traditional ranking factors are not feasible. Thus, semantic search engines try to provide concrete answers or information on the query keywords by using the semantics and relationships in their data set.

---

[*] http://hakia.com
[†] http://www.freebase.com
[‡] http://trueknowledge.com
[§] http://dbpedia.org

4.2.2.2.3   Real-Time Search Engines

The aforementioned news search engine of Google is also one example of a real-time search engine that tries to provide the user with up-to-date results matching a query. Other systems in this category are Twitter search,* Twazzup,† and Topsy.‡ While the first only uses Twitter as a source for searching new tweets, the others also include several news, weather, photo, and video websites.

To support real-time search, crawling, indexing as well as ranking need to be adapted. As Twitter search only relies on its own database, crawling is not needed. All other search engines have to crawl and index all sources of real-time content very frequently in order to provide up-to-date results. Google is monitoring more than a billion pages per day for changes [9], and OneRiot, an already closed real-time search engine, needed only 0.8 seconds to index a complete website [10]. Hence, speed and plentiful resources are important factors when indexing dynamic content.

Ranking real-time content is another issue, as it needs to be adapted to the characteristics of recently added resources. Some of the traditional ranking factors such as number of links pointing to the new resource cannot be harnessed due to their short age and consequently missing links. Others, such as keyword frequency or text quality, are not applicable for some kind of resources like tweets due to their limitation to 140 characters. OneRiot used 26 factors [9] to rank the results at search time, of which four are revealed in an online article [10]. First, as real-time results are the key point of their search engine, they use the *freshness* of the result: Recent data is more interesting than older information. The second, related aspect is *hotness*, i.e., a measure of whether the popularity of a resource is increasing or decreasing. In Twitter, for example, this can be measured by the number of retweets, i.e., if other people share the same tweet. The last two factors are about the reputation of domains respectively persons. News items published by a trusted website are ranked higher than items published in a private blog. Likewise, tweets published by a person with many followers, i.e., people following the stream of tweets of a Twitter user, are ranked higher than postings from persons with few followers.

4.3   SEARCH IN SENSOR NETWORKS

Searching for sensors can be categorized by the searched information. The first type of search is concerned with static metadata of sensors. This is a description of the sensor, its capabilities, the measured unit, or, in the case of fixed sensors, the location. The second type is concerned with the recent dynamic data, i.e., the current or historical sensor output as well as aggregations over these measurements. While the sensors output low-level measurements, the user is often interested in the high-level state of a sensor, e.g., the entity carrying the sensor or being observed by it. Furthermore, the user wants to know the location or identity of the found entities.

Due to the scarce resources of sensor platforms with respect to bandwidth, memory, and energy capacity on the one hand, and due to the high update rate of sensor

---

* http://www.twitter.com
† http://www.twazzup.com
‡ http://www.topsy.com

output on the other hand, the concepts used in traditional search engines cannot be adopted directly. Even if all dynamic data could be gathered over time, indexing this vast amount of data in real time seems infeasible. We will discuss both forms of search along with problems and possible solutions and present first systems tackling the field of searching sensor networks.

4.3.1 GENERAL CONCEPTS

Sensor networks are characterized by limited resources in terms of processing power, bandwidth, memory, and energy capacity. However, they are commonly connected to the Internet via powerful gateways/proxy servers. Hence, a shift of resource-demanding tasks to the edge of the network can prolong the network's life-time. Additionally, the communication with and inside the sensor network should be reduced to a minimum.

4.3.1.1 Gathering Sensor Data

To retrieve up-to-date readings from sensors, a pull-based approach is often used in sensor networks. Whenever readings from a sensor are needed, they are actively requested by the gateway or another computer on the Internet (see Pull Request Phase of Figure 4.3a). The sensor then returns its most recent reading (see Pull Response Phase of Figure 4.3a). This is similar to the crawling concept of search engines, but instead of regular crawls, sensors are crawled on demand, when a query is posed. Due to the request- and response-phase of pull-based approaches, the time needed for collecting readings is rather high. Additionally, with a growing number of requests, wireless communication in sensor networks is the bottleneck, resulting in higher latencies. Hence, time for answering queries increases, resulting in a decrease of user search experience.

Consequently, a push-based approach, where sensors regularly push their new readings to a base station (see Periodic Push of Figure 4.3b), performs better with respect to latency because the time for requesting is omitted. However, data may already be outdated when presented to the user, depending on the reporting interval used by the sensors. Anyhow, because of the limited resources, the update interval cannot be arbitrarily reduced. Another option is to only send updates when the state of the underlying phenomenon or entity changes (see Push on Change of Figure 4.3b). This, however, is only feasible if state changes happen not too often. If sensors have to report too often due to a short interval or fast-changing measurements, congestion of the wireless channel results in delayed or missing messages. Although latency normally can be reduced with push-based approaches, it might be a waste of resources due to lack of queries for the reported measurements.

As static metadata is not changing over time, it only needs to be transmitted once to a base station. Hence, both, pull- as well as push-based approaches are feasible in terms of resource consumption.

To further reduce the data volume when transmitting dynamic sensor data, prediction-model-based approaches, where future sensor output is predicted from past readings, can be used. Here, either a base station or the sensor itself estimates the parameters of a prediction model. A base station has more processing power and

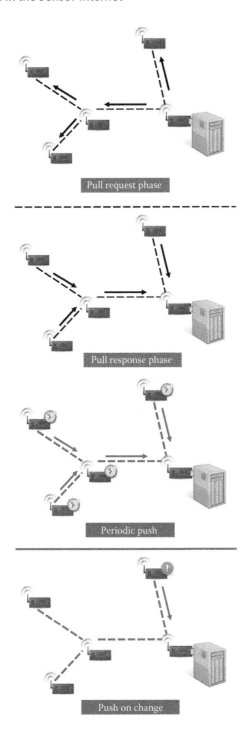

FIGURE 4.3 Example flow of pull- and push-based data gathering.

thus can compute more-complex models. However, a series of sensor readings needs to be transmitted to create an accurate prediction model. Hence, a sensor calculating model parameters itself and only transmitting these parameters may be superior to the base-station-based approach. With prediction models, latency is reduced due to the prediction at a base station instead of contacting a sensor over an unreliable (multihop) connection.

4.3.1.2   Indexing Sensor Data

To reduce data volume on the wireless channel, meta descriptions are often transmitted as keywords such as *temperature* and *celsius* or key-value pairs such as *type:temperature* and *unit:celsius*. Thus, no resource-intensive parsing and keyword extraction needs to be performed. Keywords can be inserted directly into an inverted index to produce a mapping between the keywords and the appropriate sensors.

To store current and historic sensor readings, an inverted index is not feasible due to the high frequency of new readings and the high insertion-update costs for the inverted index. Instead, sensor readings are usually stored in a database or file as a series of readings along with time stamps.

When using prediction models as the desired data-retrieval approach, only the type and parameters of the model need to be stored for each sensor instead of actual sensor output. Hence, in terms of memory usage, this is more efficient than storing the complete data stream.

4.3.1.3   Understanding User Intent

As with traditional search engines, users express their search intent by using keywords, describing the entities or sensors they would like to find. While one part of the keywords is used to describe the properties of the sensor, such as the unit, type of sensor, or accuracy, the other characterizes the entity, such as a table or an environment like a room, that is observed by the sensor.

The mapping from a sensor to the observed entity or environment is most often static. While deploying sensors or attaching them to objects, the meta description is extended with terms describing the entity monitored by that sensor. As soon as this description is known to the search engine, the entity and appropriate sensor are findable.

To indicate the intent to search for a dynamic property, the syntax *scope:value*—often used in traditional search engines today, to restrict, for example, the filetype—is adopted. The *scope* characterizes the dynamic property such as the occupancy of a room or the temperature measured by a sensor, and the *value* specifies the desired state of the property.

4.3.1.4   Presenting Results

As the number of sensors in the Internet of Things (IoT) will rise to billions in the next decades, it is important to find an appropriate subset of sensors for a query. Especially when it comes to finding sensors with a certain state, restricting the number of results (and thus the sensors that need to be contacted) is essential to account for the limited resources. Hence, most systems to allow defining or prescribing a value k as the number of results to return.

Ranking in recent systems relies only on the number of matching keywords, the current state, or the distance from the user. It is likely that as soon as sensor search engines gain popularity, untrusted parties will try to spam the results. Hence, additional factors such as trustworthiness will be included in future systems. Since limited resources are an important aspect that must be considered, it is conceivable that residual energy or latency will be other factors to be integrated.

Once the k best-matching sensors or entities have been found, they are ranked and presented to the user. Usually the meta description of the sensors and/or entities is shown. To locate the sensor, either a textual description of the place is given, or sensors are depicted on a map-based interface.

4.3.2 SYSTEMS

We now present selected search systems for sensor networks. These systems adopt the ideas discussed previously to provide resource-efficient discovery of sensors. While most presented systems are built for finding sensors matching a static keyword description, there are only a few supporting the search for sensor output in real time.

4.3.2.1 Snoogle/Microsearch

This system [11, 12] uses a two-tiered architecture (see Figure 4.4) to provide a keyword-based search to find and locate real-world entities. The lower tier consists of sensor nodes called object nodes and so-called Index Points (*IP*). Object nodes are attached to physical objects like books or folders and contain keywords describing these objects. While object nodes are mobile due to the mobility of their physical objects, the *IP*s are static sensor nodes managing object nodes in the vicinity. On the upper tier, a Key Index Point (*KeyIP*) maintains the data of several *IP*s in the network.

Because of the mobility of objects, attached object sensors can leave or join the area covered by an *IP*. To keep information at every *IP* up to date, two different modes are supported. In the default mode, the *timer* mode, object sensors periodically push

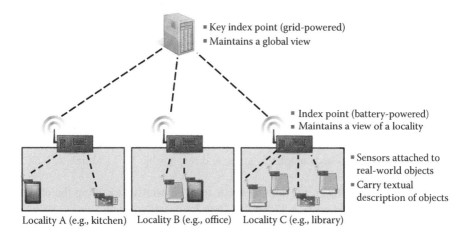

FIGURE 4.4 Two-tiered architecture of Snoogle.

their data to the next *IP*. In contrast, in *beacon* mode, the *IP* periodically pulls the textual description from the sensors by broadcasting a request and collecting the responses from object sensors.

As soon as an *IP* has collected data from the object sensors, it builds an inverted index for fast lookup of matching object sensors for a query. This aggregated information is stored at the *IP* and additionally forwarded to the *KeyIP*. That is, the *KeyIP* has an aggregated view of all *IPs* it is responsible for. Users can query the system with one or more keywords using a mobile device such as a smart phone.

There are two ways of querying Snoogle. In the local query mode, an *IP* is queried directly, returning *k* matching object sensors managed by this *IP*, i.e., only object sensors in the vicinity of the *IP*. These *k* sensors are ranked according to the number of matching keywords.

In global search mode, the query is posed to the *KeyIP* to find the top *k*-ranked sensors among all *IPs*. Instead of using the naive approach of retrieving a list of *k* sensors from each *IP* and computing the global ranked list at the *KeyIP*, thus wasting a lot of resources, Snoogle uses an incremental algorithm. First, the *KeyIP* requests from every *IP* its top-ranked sensor and creates a new ranked list with these sensors. This list is filled incrementally by contacting the associated *IP* of the ranked sensors in descending order. Thus, first, the *IP* of the highest ranked sensor is contacted with the weight of the second-ranked sensor. The *IP* returns all sensors with a weight between the given weight and its own top-ranked sensor, which was already sent in the initial step. This list of sensors is merged into the *KeyIP*'s list. Afterward, the *IP* of the second sensor is contacted with the weight of the third sensor of the initially created list, and so on, until the merged part of the list contains *k* sensors.

The two main drawbacks of the system are as follows: First, the search is limited to static textual descriptions of object sensors. Although it would be possible to use the update mechanism of *timer* or *beacon* mode to support search for dynamic data, every update would need a communication between object sensor and *IP*, which does not scale. The second limitation is another scaling problem due to the centralized nature of the *KeyIP* on the upper tier, which manages the whole network by maintaining a global aggregated view.

4.3.2.2 MAX

Similar to Snoogle, MAX [13] allows the user to search for lost objects. In the three-tiered architecture, depicted in Figure 4.5, the lowest tier consists of RFID-tag-equipped mobile objects such as books, tablets, and USB sticks. These tags contain descriptions about the object they belong to. Middle-tier *substations* are mainly static objects such as couches and tables. Sensor nodes are attached to these objects and are able to communicate with the RFID-tags and the upper-tier. The *base stations* on the upper tier are completely static and responsible for localities such as a living room or library.

To search for an object, the user needs to enter keywords describing the object one is looking for. The query is forwarded from base stations to all substations and further to the tags in communication range. Tags count how many of the keywords match their own description. This information is pulled by the substations and propagated to the base stations, which return a list of matching tags to the user ranked by

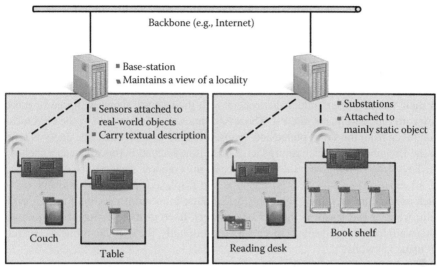

FIGURE 4.5   Three-tiered architecture of MAX.

the number of matched keywords. For the top k tags, the full descriptions, pulled by the base stations from the tag, are provided to the user. To estimate the location of an object, tags are associated to the substation, which measures the highest received signal strength indication (RSSI) value of a tag, i.e., the power present in the received wireless signal. Due to the knowledge of locations of substations, the user can easily find the lost object.

Although base stations can be connected via backbone such that the system can be enlarged to several localities, the approach of propagating each query down the hierarchy to all substations and tags hinders scaling. Similar to Snoogle, only search for static information of objects is supported. However, if tags are exchanged by sensing devices, dynamic information also could be queried due to the fact that every lower-tier object is contacted for each query.

4.3.2.3   Sensor Directories
In the past few years, several sensor data-hosting platforms such as SenseWeb,[*] Xively[†] and Paraimpu[‡] have been launched, allowing to publish live and historical sensor data on the Web. This allows access to measurements of thousand of sensors to everybody. Due to application process instructions (APIs) provided by some platforms, developers and researchers can build new systems and mashups on top of sensor data streams.

---

[*] http://www.sensormap.org
[†] Xively (https://xively.com) was first known as Pachube (http://pachube.com) then renamed Cosm (http://cosm.com). It still provides the described functions.
[‡] http://paraimpu.crs4.it

Pachube allows registering sensors and hosting the data stream provided by the sensor. Along with the position of the sensor, one can optionally specify the unit of the sensor, tags describing the sensor, whether it is indoors or outdoors, and whether it is fixed or mobile. Sensors can either push their data via an HTTP *Put* request, or they can register to be pulled periodically by Pachube. The approach of Kamilaris et al. [14] uses Pachube to allow users to search for sensors matching given keywords in their vicinity. Users query the system with their smart phone with one or more keywords and a radius of interest. The area of interest is determined by the GPS location taken from the smart phone and the user-defined given radius. The query is send to the Pachube API, which returns a list of sensors located in this area and matching the keyword(s). Along with the sensors, basic information, such as type and distance of all sensors, is retrieved from Pachube and displayed. Additional information—such as the current sensor value; historical data; minimum; maximum; measuring unit; whether the sensor is indoors or outdoors, fixed or mobile—can be requested afterward for a desired subset of sensors. Additionally, the user can refine the search by using keywords.

The keyword-based refinement allows searching for sensors that match a static description. However, due to the fact that Pachube allows the publishers to describe their sensors with their own words and with as much detail as they want, many sensors lack information that would be needed to find them. Although the live data of sensors can be displayed, there is no possibility to explicitly search for sensors with a certain output.

4.3.2.4   Objects Calling Home

With the system described Frank et al. [15], the user can search for lost objects such as books or keys. To track the location of these objects, they are equipped with sensor nodes that are detected by mobile sensors via Bluetooth. In contrast to other systems, no meta description is carried on the sensors, but only the identity of the object. When posing a query, the users cannot only specify what object they are looking for, but also a timeout t and a budget in terms of a number of messages, q. The parameter t limits the time a query is installed in the system, because it might need some time until a mobile object sensor comes in range of the lost object and detects it.

To enable scalability, queries are not sent to every sensor, as in other systems. Different heuristics are used to distribute the query to locations where the lost object most probably resides. In the beginning, objects can be searched by everyone, but objects can be associated with a certain user by using shared key exchange to restrict search for private objects to this user. Additionally, this association is used for two heuristics. First, a query is disseminated to areas where the user spent much time because the likelihood of losing an object there is much higher than in rarely visited areas. Second, users often search for objects they lost not long ago, making it more probable for the objects to be discovered at a location visited by the user in the recent past. Thus, queries are broadcasted to recently visited regions. The third heuristic is independent of the association of the user and only takes into account the last known position of a lost object. It is likely for the object to be near this position.

The user sends a query to a central server, which then selects different object sensors near the location that is computed by the heuristics. The user can also give different weights to the heuristics to prioritize one heuristic over another. A continuous query is installed at the selected object sensors. As soon as an object sensor detects the presence of the lost object, it reports this to the central server. In the case where the user specified q or t, it might happen that an object will not be found either because the query is uninstalled after the specified time or after the number of messages is exhausted without finding the object.

Although the system does not address search for static or dynamic data, but only for objects themselves, it can be mapped to the problem of searching for sensors with a certain output. Consider the case where object sensors output the identity of tagged objects in range as their sensor "measurement." The problem then is to find object sensors having this identifier as their current "measurement."

Due to the query dissemination with the help of heuristics, the system can scale to large networks. Unfortunately, there is no guarantee of finding a specific object if a budget q or a maximum time t is set. Additionally, the computation of the underlying heuristics model causes overhead, e.g., the mobile object sensors have to be tracked any time to know where detected objects are located.

4.3.2.5 Distributed Image Search

Distributed Image Search (DIS) [16] exploits camera sensors to take pictures of the environment. A user can query the system by submitting an image to it. DIS returns a list of k sensors that capture similar scenes. The list is sorted with the most relevant sensors on top, i.e., the sensor that captures the best matching image. Thus, with DIS it is possible to search for places with a certain state.

Due to the size of captured images and limited resources in sensor networks, it is infeasible to transmit the unprocessed images. Therefore, interesting characteristic features are extracted from the images at the sensor node and mapped to so-called *visterms*. These visterms are similar to keywords, describing the image. This extraction and mapping is done in two steps. First, a Scale-Invariant Feature Transform (SIFT) algorithm maps the image to continuous 128-dimensional vectors. Because there are hundreds of these vectors for an image, they are clustered and afterward mapped to keyword-like visterms describing the image. For this purpose, a precomputed dictionary has to be stored in the flash memory of the sensor nodes.

The computed visterms are used to estimate the similarity between two images. The more visterms two images have in common, the more similar they are. For fast lookup, an inverted index of visterms is maintained for all captured images at a sensor node.

There are two different search modes. In the centralized approach, all sensors push their visterms to a base station which then indexes the visterms, resulting in a centralized index of all captured images. The second approach is a local search where each sensor provides an interface for search. Visterms are pushed by the base stations to the sensors, which then return a list of top k best matches. The results of all sensors are merged at the base stations, and for the resulting top k matches, the images are retrieved from the sensors.

As with other systems, DIS does not scale to large networks because every query needs to be sent to each sensor. Another drawback is the approximate matching of extracted image features and the need for the user to query the system with an image describing the sought scene, which might not be easy to obtain. Resource consumption in terms of memory is rather high compared to other systems due to the need to store the dictionary and to maintain an inverted index.

4.3.2.6 Presto

Presto [17] allows querying a proxy for the measurement of sensors inside the sensor network managed by this proxy. In the ideal case, the query does not need to be pushed down the sensor network, but can be answered by the proxy itself. This is done by maintaining a prediction model for every sensor at the proxy to estimate the current value of the sensor.

The two-tier architecture consists of sensors on the lower tier connected to the Presto proxy on the upper tier. In the initial phase, the proxy collects data from all sensors and generates time-series-based prediction models for each sensor. The models have a confidence interval representing the error in predicted values. The proxy then sends models and its parameters to the respective sensors.

Each sensor periodically takes measurements and checks them against its own prediction model. If a sensor reading deviates more than a given threshold, it is pushed to the proxy to inform it of misprediction.

A query to the system consists of the sensor of interest and an error bound by which the query result may deviate from the true value. If the error bound is lower than the confidence interval of the associated model, the prediction is used to answer the query. However, if an outlier was reported, this is used instead of the prediction. In the case of an overly tight confidence interval, the current value is pulled from the sensor.

Prediction models are updated after a defined amount of outliers have been reported by a sensor. Additionally, predicted as well as reported outliers are stored at the proxy to answer queries for historical data.

Although the system can only be queried for the measurement of a given sensor, it would be easy to modify such systems so that search for sensors with a certain state is possible. For that, all prediction models need to be executed at the proxy, and all sensors matching the sought value given by the query are reported as result. A drawback of the system is the initial amount of data that needs to be sent to the proxy to compute the prediction models.

4.3.2.7 Dyser

Dyser [18] not only allows search for static characteristics of sensors, but especially for sensors with a certain state. To avoid contacting every sensor for each query and to reduce time for presenting results, Dyser uses Sensor Ranking, a prediction-model-based discovery strategy.

Figure 4.6 shows the architecture of the system. In Dyser, sensors are connected via a sensor gateway to the Internet. Sensors monitor the state of entities, and gateways publish an HTML page for each of these entities. Along with static textual description, dynamic information from the sensors is embedded into the HTML pages via Microformats, semantically enriched HTML-compatible tags.

The dynamic parts of the entity pages are prediction models for the sensors. These models are generated from historical data by the sensor gateway. Due to their rich resources, gateways can compute complex models.

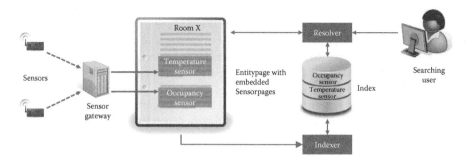

FIGURE 4.6   System architecture of Dyser. (Adapted from [18].)

Dyser supports three different types of prediction models. The simplest is an aggregated prediction model calculating the fraction of time in which a sensor had a specific output during a time window in the past, and it uses this fraction as a prediction for future outputs. The other two models try to find periodic patterns in data streams to forecast future sensor output. While the simpler model only identifies one periodic pattern, the more complex model is able to find multiple patterns with different periods.

The entity pages are crawled by Dyser, and the static and dynamic content is indexed. Afterward, all entities along with their sensors and prediction models are stored in the database.

The user can query the system with a mixture of static and dynamic keywords. For example, one can search for a free room by submitting *room occupancy:free* to the system. While *room* is the static part indicating that the user wants to find an entity of type room, *occupancy:free* is the dynamic part meaning that the occupancy of the entity should be free at the moment.

Whenever a query is posed to the system, the Resolver performs the Sensor Ranking (see Figure 4.7). First, all entities matching the static part of the query are

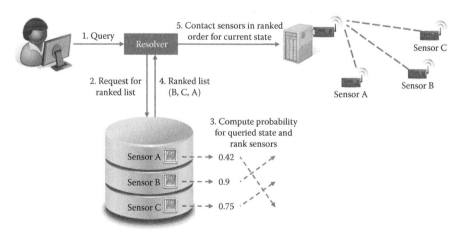

FIGURE 4.7   Sensor ranking.

retrieved along with their prediction models. By executing the models, a probability for each entity that it matches the sought state is estimated using the indexed prediction models. The sensors of the entities are sorted in decreasing order of probability. The Resolver then reads the sensors to obtain their current state, hopefully finding matching sensors first, until k matching sensors have been found or all sensors have been contacted.

The use of Sensor Ranking can reduce the number of communications with sensors substantially. However, the success is strongly dependent on the predictability of the sensor readings. If the prediction models cannot estimate the likeliness for a state sufficiently well, sensors are contacted in arbitrary order, thus leading to no advantage compared to the baseline of contacting sensors randomly. Additionally, as in Presto, the prediction models are generated outside the sensor network, leading to an initial stream of readings that needs to be transmitted.

4.4 A SEMANTIC SEARCH ENGINE FOR THE SENSOR INTERNET

As outlined in the previous sections, a sensor search engine should support discovery of sensors' respective entities with given static or dynamic properties in real time while at the same time using as few resources as possible. We present our prototype of a sensor search engine exploiting semantics and prediction models to meet these requirements.

4.4.1 System Architecture and Implementation

The prototype of the Semantic Sensor Search Engine [19] allows the user to discover entities of a certain kind and/or with a certain state. The architecture of the system is shown in Figure 4.8.

4.4.1.1 Gathering Sensor Data

The system supports real sensors attached to entities as well as so-called virtual sensors. Virtual sensors allow periodically extracting interesting real-time data from Web pages, e.g., temperature from weather sites. The real sensors are integrated via 6LoWPAN/IPv6 on the network layer. Both types of sensors are accessible via Representational State Transfer (REST), a paradigm using the methods of HTTP (*Get*, *Post*, *Put*, and *Delete*) to retrieve, create, modify, or delete resources on the Web. Each resource, i.e., in our scenario, each sensor and sensor node, is uniquely identifiable by a URI. REST does not specify the format of URIs or of the returned result. Hence, we define an address scheme as well as result format that is shared by every REST-Endpoint. The real sensors use the current IETF (Internet Engineering Task Force) CoAP (Constrained Application Protocol) Draft, a lightweight alternative for HTPP in resource-constrained environments. Thus, a REST-Endpoint also acts as a protocol-translating gateway for a sensor network.

A REST-Endpoint on the one hand provides static meta information such as location, sensor type, and the type of the entity a sensor is attached to. On the other hand, it provides dynamic information, i.e., the measurements of sensors. As an additional feature, the REST-Endpoint can be used to infer high-level states from

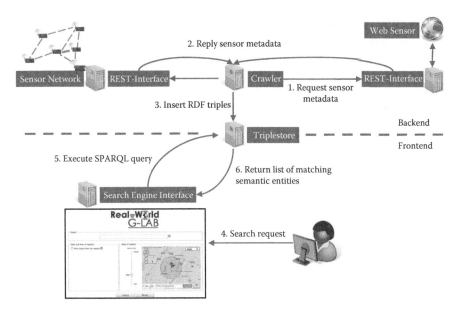

FIGURE 4.8   System architecture of the semantic sensor search engine.

raw output of sensors. For example, a detection of a passive infrared sensor in a room can be converted to an occupancy state for that room. The crawler only needs to know the addresses of the REST-Interfaces to periodically retrieve information from each endpoint.

4.4.1.2   Indexing Sensor Data

All information about sensor nodes and their sensors are stored in a triplestore. Hence, every piece of information retrieved from real or virtual sensors needs to be in the form of semantic subject-predicate-object triples. While the information provided by the real sensors already use this format, the extracted information of virtual sensors is converted into triples at the REST-Endpoint.

After each data-gathering round, the crawler updates the triplestore by inserting new triples and deleting old ones, i.e., information of sensors not present anymore at any REST-Endpoint. Thus, mobile as well as depleted sensors can be handled. Additionally, the crawler contains a component using historical values from sensors to generate prediction models for each sensor. These models are encoded as triples and stored in the triplestore, too.

4.4.1.3   Understanding User Intent

SPARQL is used to retrieve information from the triplestore. Because nonexpert users will not be willing or able to learn a query language as complex as SPARQL, we provide a graphical Web interface where keyword-based queries can be entered. Queries, consisting of static and dynamic parts as in Dyser, are converted to SPARQL queries at the cost of reduced query expressiveness.

4.4.1.4   Presenting Results

In the case of entirely static queries, information is only fetched from the triplestore, thus consuming no resources on sensor nodes. If queries for dynamic states are posed to the system, the matching entities along with their prediction models are looked up in the triplestore. As in Dyser, Sensor Ranking is executed, and sensors are contacted for their current reading in decreasing order of probability that they match the query. The results are presented in textual form, i.e., the static information and, in the case of a search for dynamic properties, the current state, and as a map with the locations of matching entities.

4.4.1.5   Future Work

When scaling the system to more sensors, the triplestore forms the bottleneck and thus needs to be improved by means of distributing it. Special data-placement policies making use of domain knowledge need to be considered to allow efficient access when scaling to billions of triples.

Further improvements can be achieved in the field of semantics. By using well-known ontologies, inference can be used to generate new facts from known ones. If the location of a sensor is tagged as a parking lot at the university and another fact states that the university is in the south of the city, a new fact can be generated allowing to discover the university parking lot when searching for a parking lot in the south of the city.

Finally, sensor nodes should be able to generate prediction models on their own to shift the burden from one instance, i.e., the crawler, to each sensor and to avoid the necessity of transmitting a series of readings for model computation outside the sensor network.

4.5   SUMMARY

Search engines for the Web have existed for nearly two decades. During that time, the techniques to crawl, index, and rank Web content have been refined to catch up with the evolving Web. This progress is continuing by means of integration of sensors and semantic data, and search engines need to support searching for this new type of information.

To this end, simple adoption of existing techniques will not be sufficient due to the heterogeneity and constrained capabilities of new device classes such as sensor networks and smart phones that are being integrated into the future Internet.

We have discussed different approaches to overcome problems when searching in sensor networks and have surveyed systems that implement these approaches. Each of these systems has its advantages and disadvantages. Hence, we combined ideas from different existing approaches and integrated them in a new architecture that can be seen as a step toward a search engine for the Internet of Things. Although it integrates techniques for energy-efficient discovery of sensors and semantic descriptions for sensors and their perceived data, further research is needed to make this approach scalable to the size of the Internet while conserving the limited resources of sensor systems to provide a good user experience in terms of latency, accuracy, and usability.

REFERENCES

1. Broder, A. 2002. A taxonomy of Web search. *SIGIR Forum* 36 (2): 3–10.
2. Gan, Q., J. Attenberg, A. Markowetz, and T. Suel. 2008. Analysis of geographic queries in a search engine log. In *Proceedings of the First International Workshop on Location and the Web (LOCWEB '08)*, 49–56. New York: ACM Press.
3. Rose, D. E., and D. Levinson. 2004. Understanding user goals in Web search. In *Proceedings of the 13th International Conference on World Wide Web* (WWW '04), 13–19. New York: ACM Press.
4. SEOmoz. 2011. Search engine ranking factors 2011. http://www.seomoz.org/article/search-ranking-factors.
5. Page, L., S. Brin, R. Motwani, and T. Winograd. 1999. The PageRank citation ranking: Bringing order to the Web. Technical Report 1999-66. Stanford InfoLab. http://ilpubs.stanford.edu:8090/422/.
6. Yandex. 2011. Spectrum. http://company.yandex.com/technologies/spectrum.xml.
7. Yandex. 2011. Matrixnet: New level of search quality. http://company.yandex.com/technologies/matrixnet.xml.
8. Hakia. 2011. Hakia search engine, 2011. http://company.hakia.com/new/whatis.html.
9. Geer, D. 2010. Is it really time for real-time search? *Computer* 43 (3): 16–19. http://doi.ieeecomputersociety.org/10.1109/MC.2010.73.
10. OneRiot Team. 2009. The inner workings of a realtime search engine: Thoughts on realtime search. http://www.docstoc.com/docs/16947406/OneRiot- Inner-Workings- of-a-Realtime-Search-Engine.
11. Wang, H., C. C. Tan, and Q. Li. 2010. Snoogle: A search engine for pervasive environments. *IEEE Trans. Parallel Distrib. Syst.* 21 (8): 1188–1202. http://dx.doi.org/10.1109/TPDS.2009.145.
12. Tan, C. C., B. Sheng, H. Wang, and Q. Li. 2010. Microsearch: A search engine for embedded devices used in pervasive computing. *ACM Trans. Embed. Comput. Syst.* 9 (4): 43. http://doi.acm.org/10.1145/1721695.1721709.
13. Yap, K.-K, V. Srinivasan, and M. Motani. 2005. Max: Human-centric search of the physical world. In *Proceedings of the 3rd International Conference on Embedded Networked Sensor Systems* (SenSys 2005), ed. J. Redi, H. Balakrishnan, and F. Zhao, 166–79. New York: ACM Press.
14. Kamilaris, A., N. Iannarilli, V. Trifa, and A. Pitsillides. 2010. Bridging the mobile Web and the Web of Things in urban environments. Paper presented at First Urban Internet of Things (UrbanIOT 2010) Workshop. Tokyo, Japan, November.
15. Frank, C., P. Bolliger, C. Roduner, and W. Kellerer. 2007. Objects calling home: Locating objects using mobile phones. In *Proceedings of the 5th international conference on Pervasive computing*, 351–68, Berlin, Heidelberg: Springer-Verlag. http://dl.acm.org/citation.cfm?id=1758183.
16. Manmatha, R., T. Yan, and D. Ganesan. 2008. Distributed image search in camera sensor networks. In *Proceedings of the 6th International Conference on Embedded Network Systems* (SenSys 2008), ed. T. F. Abdelzaher, M. Martonosi, and A. Wolisz, 155–68. New York: ACM Press. http://lass.cs.umass.edu/~yan/pubs/SenSys08-imgSearch.pdf.
17. Li, M., D. Ganesan, and P. Shenoy. 2009. PRESTO: Feedback-driven data management in sensor networks. *IEEE/ACM Trans. Netw.* 17 (4): 1256–69. http://dx.doi.org/10.1109/TNET.2008.2006818.

18. Elahi, B. M., K. Römer, B. Ostermaier, M. Fahrmair, and W. Kellerer. 2009. Sensor ranking: A primitive for efficient content-based sensor search. In *Proceedings of the 2009 Intl. Conference on Information Processing in Sensor Networks*, 217–28. Washington, DC: IEEE Computer Society.
19. Pfisterer, D., K. Römer, D. Bimschas, O. Kleine, R. Mietz, C. Truong, H. Hasemann, et al., 2011. Spitfire: Towards a semantic Web of Things. *IEEE Communications Mag.* 49 (11): 40–48.

Section II

Networking Protocols

5 Traffic Management in Wireless Sensor Networks

Swastik Brahma, Mainak Chatterjee,
and Shamik Sengupta

CONTENTS

In this chapter, we propose a distributed congestion control algorithm for tree-based communications in wireless sensor networks that seeks to adaptively assign a *fair* and *efficient* transmission rate to each node. In our algorithm, each node monitors its aggregate output and input traffic rate. Based on the difference of the two, a node then decides to increase (if the output rate is more) or decrease (if the input rate is more) the bandwidth allocable to a flow originating from itself and to those being routed through it. Since the application requirements in sensor network follow no common trait, our design *abstracts* the notion of fairness, allowing for the development of a generic utility-controlling module. Such separation of the utility- and fairness-controlling modules enable each one to use a separate control law, thereby portraying a more flexible design. The working of our congestion control is independent of the underlying routing algorithm and is designed to adapt to changes in the underlying routing topology. We evaluate the performance of the algorithm via extensive simulations using an event-driven packet-level simulator. The results suggest that the proposed protocol acquires a significantly high goodput of around 95% of the actual transmission rate, converges quickly to the optimal rate, and attains the desired fairness.

5.1 INTRODUCTION

Transmission control protocols are a key enabling technology in many of today's sensor network applications. Applications such as habitat monitoring [1], structural health monitoring [2–4], and image sensing [5] are high-data-rate applications that heavily rely on congestion control techniques, which are an integral part of any transmission control protocol [6]. Well-designed congestion control techniques allow efficient transmission of significant volumes of data from a large number of nodes along one or more routes toward the data-processing centers (usually known as a *sink* in sensor network terminology). In these high-data-rate applications, bulk data is often generated in addition to the constantly sensed data. For example, in structural health monitoring, a set of sensors is deployed in a civil structure such as a building. Each sensor measures structural vibrations continuously and transmits the data to the sink at a certain rate. When the sensors detect a significant anomaly, they generate and send out volumes of data at a much higher rate. Without congestion control, under such traffic characteristics, network collapse due to congestion is inevitable.

A common symptom of congestion in sensor networks is the increase in buffer drop rate and packet delay, similar to traditional wired networks. In addition to these, another key result of congestion in wireless sensor networks is the degradation of radio channel quality. This is because concurrent data transmissions over different radio links interfere with each other, causing channel quality to depend not just on noise, but also on traffic densities. Moreover, the time-varying nature of the radio channel and asymmetry in communication links make it harder for even regulated traffic to reach the sink. Another consequence of congestion in multihop wireless sensor networks is that the network gets biased toward delivering data from nodes nearer to the base station and is grossly unfair toward nodes farther away.

Congestion also has deleterious effects on energy efficiency. As the offered traffic load crosses a certain point of congestion, the number of bits that can be sent with the same amount of energy decreases. The network ends up wasting energy by transmitting packets from nodes upstream in the network toward the sink, only to be dropped. Thus the goal is not only to achieve high channel utilization, but also to communicate efficiently in terms of energy consumed per unit of successful communication.

Furthermore, providing fairness among flows is also highly desirable. However, the notion of fairness is hard to ground in sensor networks, since the application requirements follow no common trait. Fairness is concerned with the relative throughput of the flows sharing a link. However, the exact apportioning of the available bandwidth among the flows is dictated by the requirements of the application. Thus, it would be beneficial if the notion of fairness could be abstracted while designing a congestion control scheme for sensor networks. This would facilitate decoupling the control of *network utilization* from the management of *fairness among different flows.* Such a separation would allow easier implementation of different fairness models according to the needs of the application at hand without considering the design of the module controlling utilization of the network. Moreover, such a design would also be beneficial for sensor networks that can support multiple concurrent applications as well as systems with multiple users [7], where there is a need to treat flows from different applications or users in an inequitable manner. In general, decoupling of the two modules simplifies the design and analysis of each one by reducing the requirements imposed, and enables redesigning one of the modules without considering the other. Such separation of the utility and the fairness controllers has been proposed for traditional wired networks by Katabi, Handley, and Rohrs [8], but designing such a scheme for wireless sensor networks is not a trivial extension of previous work.

In this chapter, we propose a distributed and adaptive mechanism for controlling congestion in sensor networks that seeks to find an optimal transmission rate for the nodes—a solution that is both fair and maximally efficient. In our scheme, we use two separate modules to control the *utility* of the network and *fairness* among flows. Each node monitors its aggregate input and output rates. Based on the discrepancy between the two, a node first computes the required aggregate increase (if the output rate is more) or decrease (if the input rate is more) in traffic. Next, the fairness-controlling module acts on this required aggregate change and apportions it into individual flows to achieve the desired fairness. Note that the utility-controlling module is completely indifferent as to how the total change in required traffic is distributed among the flows. Now, to communicate the efficient transmission rate of a flow as calculated by the intermediate nodes along its path to the flow's source, we piggyback control information into data packets and further utilize the broadcast nature of the wireless medium. This enables a child node to overhear the bandwidth allocated to a flow by its parent. The child node, on its part, also computes the bandwidth that can be allocated to the flow by itself, and then compares this with the bandwidth allocated to the flow downstream and propagates the smaller rate upstream. The proposed protocol adjusts its aggressiveness according to the spare bandwidth in the network. This reduces oscillations, provides stability, and ensures efficient utilization of network resources.

Decoupling the utility- and fairness-controlling modules in our protocol opens new avenues for service differentiation using schemes that provide desired bandwidth apportioning. The scheme also supports multiple concurrent applications, is highly robust to changes in the underlying topology and routing dynamics, and can adapt to changes in them. Moreover, the protocol also has an additional advantage of improving channel quality by inducing a *phase-shifting effect* among neighboring nodes, which introduces a slight jitter to the periodicity of the application, thereby breaking the synchronization among periodic streams of traffic.

We analyze the performance of the algorithm via extensive simulations using an event-driven packet-level simulator written in Java. For simplicity, we build our work upon the sensor network de facto standard MAC layer, CSMA (Carrier Sense Multiple Access), and link-quality-based path selection. Our simulation results suggest that the steady-state behavior of the protocol can acquire a goodput of around

95% of the actual transmission rate for all nodes, achieve the optimal transmission rate quickly, and furthermore attain fairness among all nodes.

The rest of the chapter is organized as follows. Section 5.2 provides a brief literature review of related works in this area. Section 5.3 formally defines the problem being looked at and also lays down the design rationale. The congestion control algorithm is presented in Section 5.4. Section 5.5 analyzes the steady-state performance of the algorithm via extensive simulations. Finally, Section 5.6 concludes the chapter.

5.2 RELATED RESEARCH

Prior work in the literature regarding sensor networks has broadly looked at two qualitatively different problems: *congestion mitigation* and *congestion control*. In general, congestion mitigation looks at the following problem: If the nodes in a sensor network are provisioned to sense the environment and send periodic samples at a fixed rate, then—when the aggregate traffic exceeds the network capacity—how should the nodes regulate their transmissions so that the network goodput and fairness degrade gracefully? This is different from congestion control, which seeks to find an optimal fair rate for the sensor nodes that is also maximally efficient. In this case, when the nodes transmit data at the optimal rate, the network is maximally utilized, and the per-node goodput is close to the sending rate.

Adaptive Rate Control (ARC) [9] monitors the injection of packets into the traffic stream as well as route-through traffic. Each node estimates the number of upstream nodes, and the bandwidth is split proportionally between route-through and locally generated traffic, with preference given to the former. The resulting bandwidth allocated to each node is thus approximately fair. Also, reduction in the transmission rate of route-through traffic has a backpressure effect on upstream nodes, which in turn can reduce their transmission rates.

Wan, Eisenman, and Campbell [10] have proposed Congestion Detection and Avoidance (CODA). CODA uses several mechanisms to alleviate congestion. In *open-loop hop-by-hop back pressure*, when a node experiences congestion, it broadcasts back-pressure messages upstream toward the source nodes, informing them of the need to reduce their sending rates. In *closed-loop multisource regulation*, the sink asserts congestion control over multiple sources. Acknowledgments (ACKs) are required by the sources to determine their sending rates when traffic load exceeds the channel capacity. In general, open-loop control is more appropriate for transient congestion, while closed-loop control is better for persistent congestion.

Sankarasubramaniam, Akan, and Akyildiz [11] have proposed the event-to-sink reliable transport (ESRT) protocol. ESRT allocates transmission rates to sensors such that an application-defined number of sensor readings are received at the sink while ensuring that the network is uncongested. The rate allocation is centrally computed at the base station. ESRT monitors the local buffer level of sensor nodes and sets a congestion-notification bit in the packets that it forwards to the sink if the buffer overflows. If a sink receives a packet with the congestion-notification bit set, it infers congestion and broadcasts a control signal informing all sources to reduce their common reporting frequency. However, this approach suffers from a few drawbacks.

Firstly, since the sink must broadcast this control signal at a high energy to allow all the sources to hear it, an ongoing event transmission can be disrupted by this high-powered congestion signal. Moreover, rate regulating all sources, as proposed in the literature [11], is fine for homogeneous applications, where all sensors in the network have the same reporting rate, but it is not fine for heterogeneous ones. Even with a network where all the sources have a uniform reporting rate, ESRT always regulates all sources, regardless of where the hot spot occurs in the sensor network. The control law used by ESRT is based on empirically derived regions of operation, and does not attempt to find a fair and efficient rate allocation for the nodes.

Fusion [12] is a congestion-mitigation technique that uses queue lengths to detect congestion. Fusion uses three different techniques to alleviate congestion: hop-by-hop flow control, rate-limiting control, and a prioritized MAC. Hop-by-hop flow control prevents nodes from transmitting if their packets are only destined to be dropped downstream due to insufficient buffer spaces. Rate-limiting control meters the traffic being admitted into the network to prevent unfairness toward sources far away from the sink. A prioritized MAC ensures that congested nodes receive prioritized access to the channel, allowing output queues to drain. Fusion focuses on congestion mitigation and does not seek to find an optimal transmission rate for the nodes that is both fair and efficient.

Rangwala et al. [13] proposed the Interference Aware Fair Rate Control protocol (IFRC). IFRC is a distributed rate-allocation scheme that uses queue sizes to detect congestion and then further shares the congestion state through overhearing. Congestion Control and Fairness for Many-to-One Routing in Sensor Networks [14] is another rate-allocation scheme that uses different mechanism than IFRC. Both IFRC [13] and the approach of Ee and Bajcsy [14] are tangentially related to our work in the sense that they attempt to find an optimal transmission rate for all the nodes that avoids congestion collapse. Note that our algorithm has greater flexibility than IFRC [13] and the congestion-control-fairness method [14], since many different traffic-allocation policies can be implemented in our congestion control scheme without changing the basic congestion control module (the utility controller). Moreover, IFRC suffers from the additional drawback of having sophisticated parameter tuning for stability, unlike ours.

Paek and Govindam [7] have proposed the Rate Controlled Reliable Transport protocol (RCRT). This protocol is built for loss-intolerant applications that require reliable transport of data from the source nodes to the sink. RCRT uses end-to-end explicit-loss recovery by implementing a NACK-based scheme. Furthermore, RCRT places all congestion detection and rate adaptation functionality in the sinks, thereby producing a centralized congestion control scheme.

Michopoulos, Guan, and Phillips [15] have proposed a congestion control mechanism to minimize packet drop by adjusting the buffer in each node according to the transmitting downstream nodes; the algorithm automatically adjusts a node's forwarding rate to avoid packet drops due to congestion. The algorithm resolves the fairness problem by allocating equal bandwidth to the sources. Yin et al. [16] proposed a rate-based Fairness-Aware Congestion Control (FACC) protocol, which controls congestion and achieves approximately fair bandwidth allocation for different flows. Their congestion control is based on probabilistic dropping based on queue occupancy and hit frequency. Our congestion control is in contrast to these works, as it abstracts the notion of fairness, allowing it to assume different fairness models, such as weighted fairness.

Zawodniok and Jagannathan [17] proposed a hop-by-hop predictive congestion control scheme for WSNs. Their algorithm detects the onset of congestion using queue utilization and a channel-estimator algorithm that predicts the channel quality. Flow control is then achieved by a backoff interval selection scheme.

In this chapter, we focus on congestion control in sensor networks, but it will not be out of place to discuss some recent work on congestion control in wireless network in general. Warrier et al. [18] proposed a cross-layer optimization scheme for congestion control in multihop wireless networks. They implemented a differential backlog-based MAC scheduling and router-assisted back-pressure congestion control scheme using real off-shelf radios. Vannier and Gurin Lassous [19] focused on fair bandwidth sharing between end-to-end flows while maintaining an efficient overall throughput in the network. They have proposed a dynamic rate-allocation solution that is based on a simple radio-sharing model.

In the next section, we will formulate our problem and also discuss the rationale behind our solution approach.

5.3 PROBLEM DESCRIPTION AND DESIGN RATIONALE

Let us consider a set of N sensor nodes, numbered 1 through N. In the simplest version of the problem, each node has an infinite amount of data to be sent to a single base station. The nodes can originate data traffic as well as route traffic originated by other nodes. Thus, each node can act both as a *source* and a *router*. The nodes sample the environment at periodic intervals, encode the information into data packets, and send them out to a central *base station* or *sink*. Let the flow originating from Node i be f_i, and let r_i be the rate at which flow f_i is injected into the network. We seek to adaptively assign the rate r_i to flow f_i that is both fair and efficient. Note that r_i is the rate at which node i injects flow f_i into the network, and does not include the rate at which node i forwards traffic.

We assume that the sensor nodes run a contention-based MAC protocol. The default MAC in TinyOS, a widely used sensor network operating system, uses carrier sensing for collision avoidance. Our implementation is based on this MAC, though it can be extended to other MAC protocols as well, such as those that use RTS/CTS [20]. We further assume that the MAC layer provides link-layer retransmissions. Specifically, the MAC protocol used in this chapter senses the medium before trying the send a packet. If the medium is busy, the protocol performs a random backoff before attempting to send again. Collisions are detected using ACKs. If a collision occurs, retransmission of the packet is attempted. Our congestion control protocol works well in a regime where the wireless loss rate over the communication links is such that link-layer retransmissions recovers from most packet losses.

We further assume that the sensor nodes run a routing protocol [21] that builds a routing tree rooted at the base station (or sink). Figure 5.1 shows an example of such a tree. The working of our congestion control algorithm is not dependent on the exact choice of the routing tree formed. However, the performance of the algorithm

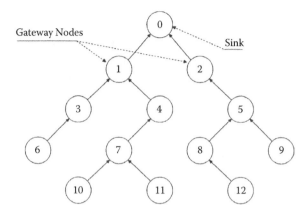

FIGURE. 5.1   An example of a routing tree.

would benefit from a tree that was formed based on link-quality metrics. In the rest of the chapter, we assume that a routing tree has been formed by the routing protocol rooted at the sink. Our congestion control algorithm is thus able to adapt to changes in the underlying routing topology.

5.3.1 DESIGN RATIONALE

A novel approach adopted in this chapter is the decoupling of the notion of *fairness* from *utility* (or efficiency) of the network. Conceptually, fairness and efficiency are independent. Efficiency involves only the aggregate traffic behavior. When the input traffic rate equals the link capacity, no queue builds up and utilization is optimal. Fairness, however, is concerned with the relative throughput of flows sharing a resource. Thus, in our model, we use two separate modules for controlling utility and fairness, viz., the *utility controlling module* and the *fairness controlling module.*

The use of two separate modules allows modification of one of the modules without redesigning the other one. For example, the fairness-controlling module can implement any fairness model without changing the underlying utility module. This is particularly attractive in sensor networks where the application requirements follow no common trait. For example, even within a typical application like temperature monitoring, while some nodes might be sending critical information such as news of a fire, others might be sending regular periodic updates. In such scenarios, the application might require more information to be received at the sink from the critical regions rather than from regions that appear safe, i.e., the application requires a higher data rate from nodes sending the critical data packets. Furthermore, as envisioned by Paek and Govindan [7], future sensor networks are likely to evolve and support multiple concurrent applications running on each mote, such as motes deployed in a building having both a temperature sensor as well as a vibration sensor. Decoupling utilization from fairness helps abstract the fairness model, and allows for the development of a generic utility-controlling module.

5.3.2 DESIGN GOALS

Our congestion control has been designed for sensor networks with high data-rate requirements, as exemplified by the applications given in Section 5.1. As mentioned previously, not only are we interested in high channel utilization, but also in communication efficiency in terms of energy consumed per unit of successful communication. Moreover, fairness is also highly desirable. The four design goals of our congestion control algorithm are as follows:

- *Network efficiency*: Our primary design goal is to operate the network at an efficient operating point. As network load increases beyond a certain congestion point, congestion collapse occurs, and no useful work gets done by the network. Nodes at the edge of the network transmit packets toward the sink, only to be dropped. This not only decreases goodput of the flows, but also wastes precious energy reserves of the nodes as energy is invested in transmitting packets only to be dropped later, thereby decreasing the number of bits that can be sent with the same amount of energy. At the same time, we need to ensure that the sensor motes send data at a sufficiently high rate, without, of course, contributing to congestion. Thus, our congestion control algorithm seeks to adaptively find a *fair* transmission rate for all nodes that is also *maximally efficient*.
- *Flexible design*: Sensor networks deliver myriad types of traffic, and the requirements of the applications vary greatly from one to another. In some homogeneous deployments, all nodes transmit data at the same rate, and it might be necessary to split up the available bandwidth equally among them. However, this is not always the case. For deployments in which some nodes require a higher bandwidth allocation (like those sending images), available bandwidth might need to be disproportionately split up among the flows sharing a resource. This calls for a need to abstract away the notion of fairness, allowing it to assume different fairness models, without changing the design of the module controlling the utilization of the network. This design is more flexible, in the sense that it allows for the development of a generic utility-controlling module, abstracting away the fairness module.
- *Support concurrent applications*: Future sensor networks are likely to support multiple concurrent applications and also evolve to become multiuser systems [7]. To allow for such evolution, it is important that the transmission control protocols provide explicit support for the same. In these systems, it might very well be necessary to treat traffic from different applications in a different manner. Abstracting the fairness module inherently facilitates such differentiation by allowing the fairness component to assume any fairness model, where available bandwidth can be disproportionately apportioned among competing flows.
- *Robustness*: Finally, it is important that any congestion control be robust to changes in underlying routing dynamics. Moreover, the algorithm also needs to adapt to changes in network topology that may be caused by nodes entering or leaving the system. For example, more nodes may be deployed,

or existing ones may die due to exhaustion of their energy reserve. All of these may require the rate allocations of the nodes to be changed dynamically without requiring external intervention.

5.4   CONGESTION CONTROL DESIGN

In this section, we develop a simple distributed algorithm for congestion control in sensor networks. Before we begin, we first define a few terms.

Feedback delay: Feedback delay of a flow is the time interval, measured starting from the time node i starts transmitting flow f_i at a rate r_a to the time the control signal arrives and changes the rate to r_b.

Gateway node: A node is called a gateway node if it is one hop away from the sink (see Figure 5.1).

Control interval: Control interval is the time period over which a node takes a control decision regarding the increase or decrease in the transmission rates of the flows originated by itself and of the flows being routed through it. In our implementation, this interval has been taken to be the average feedback delay of the flows passing through the gateway node.

5.4.1   OVERVIEW

Each node i, which originates flow f_i, maintains and sends (a) the current rate r_i at which the flow is being injected into the network and (b) an estimate of the *feedback delay* of the flow by appending a congestion header to each packet being sent in the flow. Figure 5.2 shows the congestion header. This header allows the intermediate

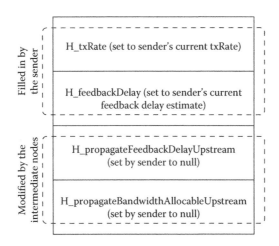

FIGURE. 5.2   Congestion header.

nodes to be aware of the transmission rates of the flows passing through it as well as their feedback delays.

In our mechanism, the nodes monitor their input traffic rate. Based on the difference between the link capacity and the input traffic rate, the nodes compute the desired increase or decrease in the rate of origination of the flows sharing the link. Note that the bandwidth that can be acquired by a flow f_i, originating from node i, is constrained by the *minimum* bandwidth that can be allocated to the flow by the nodes along the path of the flow. Thus, the effective transmission rate of f_i that needs to be conveyed to node i is the minimum transmission rate that can be sustained by the nodes along the path of the flow. To enable this, we take advantage of the broadcast nature of the wireless medium by assuming that the transmission of a packet by a parent node can be heard by its children. When a node (say j) transmits a packet belonging to flow f_i, it writes into the packet header the minimum of the bandwidths that can be allocated to f_i by itself and by the nodes downstream (nearer to the sink) to node j. Node j knows the minimum bandwidth allocated to f_i by its downstream nodes by overhearing transmission of its parent. Finally, the congestion signal that reaches the source node is, thus, the minimum bandwidth allocable to the flow based on the spare bandwidths of the nodes along the path of the flow. Next, we formally present the congestion control algorithm.

5.4.2 CONGESTION CONTROL ALGORITHM

The congestion control algorithm can be described by the following steps executed at each node at every control interval:

1. Measure the average rate r_{out} at which packets can be sent from the node, the average aggregate input rate r_{in}, and Q, which is the minimum number of packets in the output queue seen by an arriving packet in a control interval.
2. Based on the difference between r_{out} and r_{in}, and Q, compute Δr, which is the total change in aggregate traffic required as: $\Delta r = \alpha \times (r_{out} - r_{in}) - \beta \times (\frac{Q}{t_{CI}})$, where α and β are constants, and t_{CI} is the control interval.
3. Apportion Δr into individual flows to achieve fairness.
4. Compare the bandwidth computed for each flow with the bandwidths advertised by its parent. Use and propagate the smaller rate upstream.

An important point that needs to be clarified is when the congestion control procedure gets invoked at the various nodes. The congestion control algorithm gets invoked at every control interval at the gateway nodes. For any other node, the algorithm is invoked when the transmission rate of its parent changes. Note that this essentially means that the congestion control must be invoked at *all* the nodes at every control interval. We will elaborate further on the rationale behind such a control behavior and on the estimation of the control interval in Section 5.4.2.6.

The congestion control algorithm outlined here requires:

1. Estimation of average aggregate output rate
2. Estimation of average aggregate input rate

3. Computation of the total change in aggregate traffic required (Δr) to control efficiency
4. Apportioning Δr into individual flows to obtain desired fairness
5. Propagation of rate upstream
6. Estimation of the control interval.

We now explain in detail the working of each one of these requirements in the following subsections.

5.4.2.1   Estimation of Average Output Rate

Let t_{out} (measured in seconds) be the time required to transmit a packet, measured starting from the time the packet was sent by the network layer to the MAC layer to the time when the MAC layer notifies the network layer that the packet was successfully transmitted. This is shown in Figure 5.3, where we assume that the MAC protocol in use is CSMA/CA (CSMA with Collision Avoidance) with an acknowledgment-based scheme. Then, we note that the effective rate r_{out} (packets per second) is the inverse of the time interval t_{out} (seconds), i.e., $r_{\text{out}} = \frac{1}{t_{\text{out}}}$.

The value of t_{out} obtained per packet transmitted is one particular instance of the average time taken to transmit a packet. We compute the average value by using the following exponential moving-average formula:

$$\overline{t_{\text{out}}^i} = (\alpha_{\text{out}}) \cdot T_{\text{out}} + (1 - \alpha_{\text{out}}) \cdot \overline{t_{\text{out}}^{i-1}}, \tag{5.1}$$

where $\overline{t_{\text{out}}^i}$ is the exponential moving-average value of the variable t_{out} in the ith iteration, α_{out} is the weight, and T_{out} is the current value of the variable t_{out}. Calculating $r_{\text{out}} = \frac{1}{\overline{t_{\text{out}}}}$ gives us the average rate at which packets can be transmitted from a particular node at any instant.

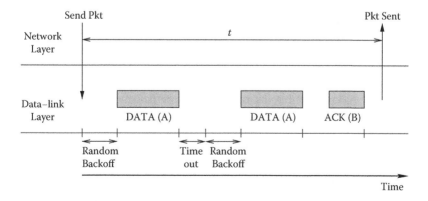

FIGURE 5.3   Determination of time t taken to successfully transmit a data packet, considering CSMA as the MAC protocol. The sender of a packet is denoted in parentheses.

5.4.2.2 Estimation of Average Input Rate

Let t_{in} (measured in seconds) be the interpacket arrival time at a node, measured starting from the time a packet was enqueued to the time when the next packet is successfully enqueued. Then, the effective aggregate input rate r_{in} at a node is the inverse of the time interval t_{in}, i.e., $r_{in} = \frac{1}{t_{in}}$. We compute the average value of t_{in} using the following exponential moving-average formula:

$$\overline{t_{in}^i} = (\alpha_{in}) \cdot T_{in} + (1 - \alpha_{in}) \cdot \overline{t_{in}^{i-1}}, \tag{5.2}$$

where $\overline{t_{in}^i}$ is the exponential moving-average value of the variable t_{in} in the ith iteration, α_{in} is the weight, and T_{in} is the current value of the variable t_{in}. Calculating $r_{in} = \frac{1}{\overline{t_{in}^i}}$ gives us the average aggregate input rate at a node at any instant.

5.4.2.3 Controlling Efficiency

In controlling efficiency, we seek to maximize link utilization while minimizing buffer drop rates and persistent queues. The efficiency-controlling component considers only the aggregate traffic and does not take into account fairness issues.

The efficiency controller computes the required increase or decrease of the aggregate transmission rate of the traffic (Δr) in a control interval (in packets per second). This is computed as

$$\Delta r = \alpha \times (r_{out} - r_{in}) - \beta \times \left(\frac{Q}{t_{CI}}\right), \tag{5.3}$$

where t_{CI} is the control interval of the node, α and β are constant parameters, and Q is the persistent queue size. We use 0.4 and 0.226 as values of α and β, respectively, based on the work done in the literature [8]. Q is computed as the minimum number of packets seen by an arriving packet in a control interval. Note that the quantity, ($r_{out} - r_{in}$), can be positive, negative, or equal to zero. When ($r_{out} - r_{in}$) > 0, the link is underutilized, and positive feedback needs to be sent to increase the transmission rates of the flows. When ($r_{out} - r_{in}$) < 0, the link is congested, and negative feedback is required to decrease the transmission rates. For the case when ($r_{out} - r_{in}$) is equal to zero, i.e., the input capacity matches the link capacity, we have to provide feedback in a manner that drains the persistent queue size Q. This is why the aggregate feedback has been made proportional to Q.

It is worth noting here that the values of the coefficients α and β are for linear feedback systems with delay. Both α and β influence the stability of the feedback system, which can be studied from their open-loop transfer function in a Nyquist plot. The values of α and β are such that the system satisfies the Nyquist stability criterion.

5.4.2.4 Controlling Fairness

The task of the fairness-controlling module is to apportion the feedback computed by the efficiency-controlling module into the individual flows. For the sake of simplicity, we treat all flows equally in the following discussion. Though there could be

other choices, we use the additive-increase/multiplicative-decrease (AIMD) algorithm to make the system converge to fairness. There are three cases:

- $\Delta r > 0$: In this case, the positive feedback to be sent is divided equally among the flows so that increase in throughput of all the flows is the same. If we assume that there are n flows comprising the route-through traffic of a node, then each one of the flows should get $1/(n + 1)$th fraction of the spare bandwidth, considering that the node also originates traffic of its own. Thus, we can write

$$r_i(t+1) = r_i(t) + \frac{\Delta r}{n+1}. \tag{5.4}$$

- $\Delta r < 0$: In this case, the negative feedback to be sent is allocated in such a way that the decrease in throughput of a flow is proportional to its current throughput. Thus,

$$r_i(t+1) = m \cdot r_i(t), \tag{5.5}$$

  where $0 < m < 1$. Here m is inversely proportional to the magnitude of Δr, $|\Delta r|$. This means that, as the discrepancy between the output rate and the input rate increases, the throughput of the flows is cut down more and more aggressively.

- $\Delta r = 0$: This case corresponds to the efficiency being near optimal. Bandwidth shuffling is done in this case such that the total traffic rate, and consequently the efficiency, does not change, yet the throughput of each individual flow changes gradually to approach the flow's fair share.

5.4.2.5 Propagation of Rate Upstream

As mentioned previously, congestion control is invoked at the *gateway nodes* at every control interval. For all the other nodes, congestion control is invoked when the transmission rate of the parent node changes. It is quite evident that this, in effect, leads to congestion control being invoked at all the nodes at every control interval. In order to propagate the congestion signal upstream, starting from the gateway nodes, we make use of the broadcast nature of the wireless medium, which enables a child node to overhear the transmissions of its parent. When a node (say j) transmits a packet of a flow f_i, it writes two pieces of information into a header (*H_propagate-BandwidthAllocableUpstream*, as shown in Figure 5.2) in the packet: the minimum of the bandwidths allocable to the flow by itself (which it computes when its congestion control is invoked) and the bandwidth advertised by its parent. Note that the bandwidth advertised by its parent is, in turn, the minimum of the bandwidth allocable to f_i by the nodes downstream to node j. Thus, the transmission rate for flow f_i that node j writes into the header of a packet of f_i is the minimum bandwidth allocable to the flow among nodes downstream starting from node j. In this fashion, the feasible transmission rate that ultimately reaches node i (which originates flow f_i) is the minimum bandwidth allocable to the flow by the intermediate nodes along the path of the flow.

5.4.2.6 Estimation of Control Interval

Control theory states that a controller must react as quickly as the dynamics of the controlled signal. Otherwise, the controller will lag behind the system being controlled and will be ineffective. Thus, the controller of our congestion control scheme makes a single control decision every average *feedback delay* period (averaged over the feedback delays of the flows passing through a node). This is motivated by the need to observe the results of the previous control decisions before attempting a new control. For example, if an intermediate node tells an upstream node to increase its transmission rate, the former should wait to see how much spare bandwidth remains before asking it to increase again.

As explained before, the feedback delay of a flow is the time interval measured starting from the time node i starts transmitting flow f_i at a rate r_a to the time the control signal arrives and changes the rate to r_b. Feedback delay of a flow is, thus, equal to the delay experienced by the packets as they flow from the source node to one of the gateway nodes plus the time taken by the congestion signal to propagate upstream from the gateway node back to the source node. Now, the time taken by the packets to travel from source nodes to the gateway nodes comprises the queuing delays of the packets plus the average time taken to transmit a packet successfully by the MAC layer at the various intermediate hops. Furthermore, the time taken by the congestion signal to propagate upstream from the gateway node to the source node comprises the summation of the interpacket arrival interval of the packets of the flow at the head of the output queue at the intermediate nodes and the time taken to transmit a packet successfully by the MAC layer at these nodes. To understand why, recall that the congestion signal propagates upstream by utilizing the broadcast nature of the wireless medium, enabling a child node to overhear the transmission of its parent. Thus, when the minimum bandwidth allocable to a flow changes at a parent node, its child comes to know about it *at the soonest* after the instantaneous interpacket arrival interval of the packets of the flow at the head of the output queue at the parent plus the time taken to transmit a packet by the MAC layer at the same.

From this discussion, it is clear that, if a flow k passes through n intermediate hops, numbered 1 through n, then the feedback delay of flow k is given by

$$\sum_{i=1}^{n} \left(queueingDelay_i^k + interPktArrInterval_i^k + 2 \cdot macPktTxTime_i^k \right), \quad (5.6)$$

where $queueingDelay_i^k$ is the average queuing delay at node i of the packets of flow k, $interPktArrInterval_i^k$ is the average time interval between two packets of flow k arriving at the head of the output queue, and $macPktTxTime_i^k$ is the average time taken by the MAC layer of node i to transmit a packet.

Now, what is required is to compute the feedback delay in a distributed manner. For this, each node keeps a moving average of each one of the three terms in

Equation (5.6) for each flow passing thorough it. When a node (say j) transmits a packet of a flow k, it writes into a header (H_propagateFeedbackDelayUpstream, as shown in Figure 5.2) in the packet the summation of the three terms as estimated by itself for flow k and the value for feedback delay for flow k as advertised by its parent. The value for feedback delay advertised by its parent is, in turn, the summation of the three terms of the nodes downstream from j. In this fashion, the source node can be made aware of the feedback delay of the flow it is originating. The source node, on its part, maintains the feedback delay experienced by its flow and sends them out to the intermediate nodes via a header (H_feedbackDelay, as shown in Figure 5.2) in every packet.

As mentioned earlier, gateway nodes are those that are one hop away from the sink. These nodes have a more holistic view of the network, since traffic from every other node has to go via one of these nodes to get to the sink. The gateway nodes use the average feedback delay of the flows passing through them as the control interval. For the other nodes in the network, the congestion control procedure is invoked when the transmission rate of its parent changes. In this scheme, effectively, the congestion control algorithm gets invoked for all the nodes for every average feedback delay of flows passing through the gateway node, which is the control interval.

5.5 PERFORMANCE EVALUATION

To evaluate the performance of the congestion control algorithm, we implemented a packet-level simulator in Java. The simulator is event-driven and implements CSMA/CA as the MAC protocol.

Table 5.1 shows the parameter values used in the simulations. We assumed that the maximum communication channel bit rate is 38.4 kbps, the ACK packet is 5 bytes, the beacon packet is 10 bytes, and data packets are 64 bytes. The simulations

TABLE 5.1
Parameter Values Used in the Simulations

Parameter	Values
Channel bit rate	38.4 kbps
Topology considered	10×10 grid
Link loss probability	0.1
Initial packet generation rate	1 pkts/s
Queue size	25
Data packet size	64 bytes
Beacon packet size	10 bytes
MAC ACK packet size	5 bytes
Tx/sensing range	25 m
DIFS	50 μs
SIFS	10 μs
Slot time	20 μs
Maximum Retx threshold	7

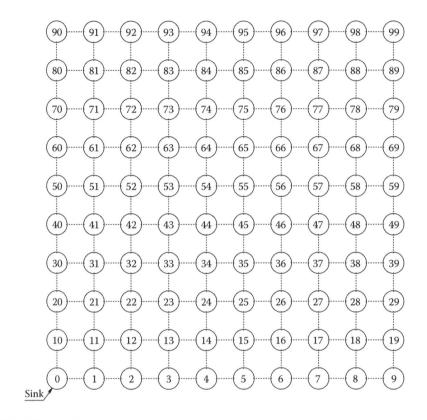

FIGURE 5.4   10 × 10 grid used in the simulations.

have been carried out on a 100-node network, a 10 × 10 grid, with the sink at the lower-left corner. The topology is shown in Figure 5.4. The nodes have been initialized with an initial packet-generation rate of 1 packets/s. Each link has a constant loss probability of 0.1, which is in addition to the loss caused by interference. Since ARQ (Automatic Repeat reQuest) is implemented, packets dropped due to interference or queue overflow are retransmitted. The weights used in the exponential moving-average calculation of the output and input rate (see Section 5.4.2.1 and Section 5.4.2.2) is set to 0.1. Each node has a queue to hold packets forwarded by its children, as well as packets generated by itself. This queue has been set to hold a maximum of 25 packets.

5.5.1   GOODPUT AND FAIRNESS

Figure 5.5 shows the per-flow goodput received at the sink. Each bar represents the average rate of transmitted packets from the corresponding node. The darker section of each bar represents the average rate at which packets from that node were received at the base station. Packet losses account for the rest (the remaining lighter section of the bar). We make an important observation from this graph. This figure validates

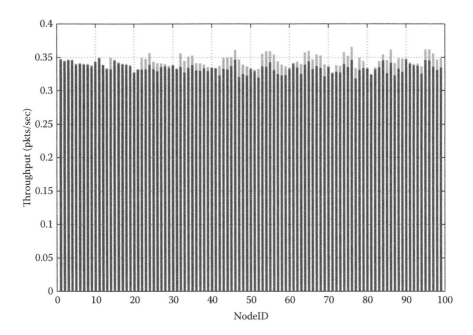

FIGURE 5.5   Per-flow goodput in the 100-node simulation.

the basic design of the congestion control algorithm, and shows that nodes receive approximately fair rates and fair goodput.

5.5.2   SUBSET OF SENDERS

Figure 5.6 shows the per-flow goodput with a subset of senders. Those nodes whose IDs are multiples of 3 were only allowed to transmit data. As can been seen from the graph, all nodes receive a fair rate. Note that the per-flow goodput is significantly higher than when all nodes transmit. This is because the protocol adapts to the increased overall available capacity and allocates it fairly to the transmitting nodes.

5.5.3   LINK-LAYER RETRANSMISSIONS

Figure 5.7 shows that without congestion control, each packet is retransmitted due to interference a maximum of about six times, a number dependent on many factors such as the topology, data packet generation rate, and size of the network. In general, the number of retransmissions, and therefore losses, increases with the depth of the network. The graph where congestion control is implemented shows a small number of retransmissions per packet, and this number roughly stays constant for all nodes in the network.

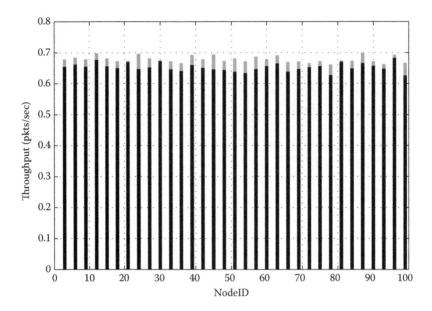

FIGURE 5.6   Per-flow goodput with only a subset of senders. Nodes whose IDs are multiples of 3 were only allowed to transmit data.

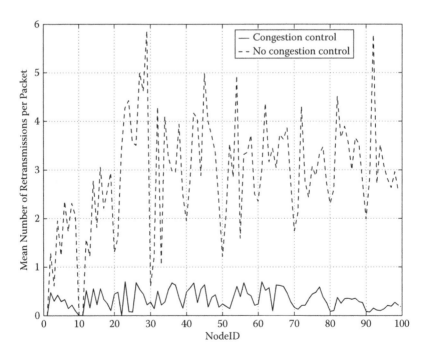

FIGURE 5.7   Mean retransmissions per node due to interference.

5.6 CONCLUSIONS

This chapter has presented a distributed algorithm for congestion control in wireless sensor networks that seeks to assign a fair and efficient rate to each node. The algorithm requires each node to monitor its aggregate input and output traffic rate, based on the difference of which the node decides to increase or decrease the transmission rates of itself and its upstream nodes. The utilization-controlling module computes the total increase or decrease in traffic rate. The fairness module decides on how exactly to apportion the total change in traffic rate required among the flows. The utilization controller is indifferent to it, thus abstracting the notion of fairness and allowing the fairness module to assume differential bandwidth-allocation policies. The congestion control is invoked at each node for every average feedback delay of the flows passing through the gateway node, which we define as the control period. By piggybacking control information on data packets and by utilizing the broadcast nature of the wireless medium, the algorithm does not require the need for additional control packets. We tested our algorithm on an event-driven packet-level simulator. The results indicate that the proposed congestion control mechanism can achieve remarkably high goodput, is able to attain fairness for all nodes in the network, and can acquire the optimal transmission rate quickly. Finally, our solution is independent of the underlying routing algorithm and can adapt to changes in the routing tree.

REFERENCES

1. Szewczyk, R., A. Mainwaring, J. Polastre, and D. Culler. 2004. An analysis of a large scale habitat monitoring. In *Proceedings of the 2nd International Conference on Embedded Networked Sensor Systems* (SenSys 2004), ed. J. A. Stankovic, A. Arora, and R. Govindan, 214–26. New York: ACM Press.
2. Kottapalli, V., A. Kiremidjian, J. P. Lynch, E. Carryer, T. Kenny, K. Law, and Y. Lei. 2003. A two-tier wireless sensor network architecture for structural health monitoring. Paper presented at SPIE 2003, San Diego, CA.
3. Mechitov, K., W. Y. Kim, G. Agha, and T. Nagayama. 2004. High-frequency distributed sensing for structure monitoring. Paper presented at First International Workshop on Networked Sensing Systems (INSS), Tokyo, Japan.
4. Paek, J., K. Chintalapudi, J. Cafferey, R. Govindan, and S. Masri. 2005. A wireless sensor network for structural health monitoring: Performance and experience. In *Workshop on Embedded Network Sensors (EmNetS)*, 1–10. Piscataway, NJ: IEEE Press.
5. Rahimi, M., R. Baer, O. I. Iroezi, J. C. Garcia, J. Warrior, D. Estrin, and M. B. Srivastava. 2005. Cyclops: in situ image sensing and interpretation in wireless sensor networks. In *Proceedings of the 3rd International Conference on Embedded Networked Sensor Systems* (SenSys 2005), ed. J. Redi, H. Balakrishnan, and F. Zhao, 192–204. New York: ACM Press.
6. Chiu, D., and R. Jain. 1989. Analysis of the increase and decrease algorithms for congestion avoidance in computer networks. In *Computer Networks and ISDN Systems* 17:1–14.
7. Paek, J., and R. Govindan. 2007. RCRT: Rate-controlled reliable transport for wireless sensor networks. In *Proceedings of the 5th International Conference on Embedded Networked Sensor Systems* (SenSys 2007), ed. S. Jha, 305–19. New York: ACM Press.

8. Katabi, D., M. Handley, and C. Rohrs. 2002. Congestion control for high bandwidth-delay product networks. In *Proceedings of the 2002 Conference on Applications, Technologies, Architectures, and Protocols for Computer Communications (SIGCOMM)*, 89–102. New York: ACM Press.

9. Woo, A., and D. Culler. 2001. A transmission control scheme for media access in sensor networks. In Seventh Annual International Conference on Mobile Computing and Networking (MobiCom), 221–35. New York: ACM Press.

10. Wan, C., S. B. Eisenman, and A. T. Campbell. 2003. CODA: Congestion detection and avoidance in sensor networks. In *Proceedings of the 1st International Conference on Embedded Networked Sensor Systems* (SenSys 2003), ed. I. F. Akyildiz, D. Estrin, D. E. Culler, and M. B. Srivastava, 266–79. New York: ACM Press.

11. Sankarasubramaniam, Y., O. Akan, and I. Akyildiz. 2003. Event-to-sink reliable transport in wireless sensor networks. *IEEE/ACM Trans. Networking* 13 (5): 1003–16.

12. Hull, B., K. Jamieson, and H. Balakrishnan. 2004. Mitigating congestion in wireless sensor networks. In *Proceedings of the 2nd International Conference on Embedded Networked Sensor Systems* (SenSys 2004), ed. J. A. Stankovic, A. Arora, and R. Govindan, 134–47. New York: ACM Press.

13. Rangwala, S., R. Gummadi, R. Govindan, and K. Psounis. 2006. Interference-aware fair rate control in wireless sensor networks. In *Proceedings of Conference on Applications, Technologies, Architectures, and Protocols for Computer Communications*, 63–74. New York: ACM Press.

14. Ee, C. T., and R. Bajcsy. 2004. Congestion control and fairness for many-to-one routing in sensor networks. In *Proceedings of the 2nd International Conference on Embedded Networked Sensor Systems* (SenSys 2004), ed. J. A. Stankovic, A. Arora, and R. Govindan, 148–61. New York: ACM Press.

15. Michopoulos, V., L. Guan, and I. Phillips. 2010. A new congestion control mechanism for WSNs. In *International Conference on Computer and Information Technology*, 709–14. Piscataway, NJ: IEEE Press.

16. Yin, X., X. Zhou, R. Huang, Y. Fang, and S. Li. 2009. A fairness-aware congestion control scheme in wireless sensor networks. *IEEE Trans. Vehicular Technology* 58 (9): 5225–34, Nov.

17. Zawodniok, M., and S. Jagannathan. 2007. Predictive congestion control protocol for wireless sensor networks. *IEEE Trans. Wireless Communications* 6 (11): 3955–63.

18. Warrier, A., S. Janakiraman, S. Ha, and I. Rhee. 2009. DiffQ: Practical differential backlog congestion control for wireless networks. In *INFOCOM Proceedings*, 262–70. Piscataway, NJ: IEEE Press.

19. Vannier, R., and I. Gurin Lassous. 2008. Towards a practical and fair rate allocation for multihop wireless networks based on a simple node model. In *11th ACM/IEEE International Conference on Modeling, Analysis and Simulation of Wireless and Mobile Systems (MSWiM)*, 23–27. New York: ACM Press.

20. Ye, W., J. Heidemann, and D. Estrin. 2002. An energy-efficient MAC protocol for wireless sensor networks. In *INFOCOM Proceedings*, 1567–76. Piscataway, NJ: IEEE Press.

21. Woo, A., T. Tong, and D. Culler. 2003. Taming the underlying challenges of reliable multihop routing in sensor networks. In *Proceedings of the 1st International Conference on Embedded Networked Sensor Systems* (SenSys 2003), ed. I. F. Akyildiz, D. Estrin, D. E. Culler, and M. B. Srivastava, 14–27. New York: ACM Press.

6 Decision-Tree Construction for Event Classification in Distributed Wearable Computers

Hassan Ghasemzadeh and Roozbeh Jafari

CONTENTS

6.1 INTRODUCTION

Wireless sensor networks have caught tremendous attention recently due to their potential for a large number of application domains. Applications range from monitoring systems such as environmental and medical monitoring to detection and supervision systems for military surveillance. A special class of these systems, called *body sensor networks* (BSNs), uses a network of lightweight embedded sensory devices to acquire and process physiological data about the subject wearing the system. BSNs can foster medical services by providing real-time and remote healthcare monitoring. They can be effective for rehabilitation [1], sports medicine [2], geriatric care [3], gait analysis [4], and detection of neurodegenerative disorders such as Alzheimer's [5], Parkinson's [6], and Huntington's [7] diseases.

Action recognition aims to detect transitional movements such as "sit to stand," "sit to lie," and "walking." Action recognition is usually required for development of many other applications. In gait analysis, certain information about the quality of gait is extracted when the person is walking. Thus, the current action needs to be reported prior to execution of other processing tasks. In monitoring Parkinson's disease (PD) patients, several symptoms such as resting tremor, muscular rigidity, bradykinesia or delayed initiation of movements, and postural instability [8] need to be detected. However, these symptoms are not equally pronounced during all of the human actions. For example, bradykinesia cannot be identified when a subject is perfectly still. In order to properly recognize PD symptoms, action recognition needs to be performed first.

Limited processing power and finite battery energy are two major obstacles in realizing real-time applications of the BSNs. The limited computing capability warrants the need for development of computationally inexpensive algorithms that run on the lightweight sensor nodes and reduce complexity of the large amount of data collected by the sensor nodes. Battery lifetime, however, can be maximized by optimizing the function of individual components such as processing unit and communication system. Studies have shown that communication consumes significantly more energy than data processing [9]. Hence, significant power savings can be obtained by enhancing the communication system.

This chapter focuses on presenting a computationally simple and distributed algorithm for action recognition. The task involves introducing a novel representation of human actions in terms of their basic building blocks, called *primitives*. With this approach, each action is represented as a set of symbols associated with the primitives. A distributed algorithm is then developed that detects human actions according to a decision-tree model. The decision tree is derived directly from interpretation of action primitives and the level of contribution of each sensor node for action identification. The distributed algorithm produces a global classification decision based on a subset of results generated by individual sensor nodes. The compact representation of actions along with the distributed nature of the algorithm enables a wearable monitoring system to lower the amount of information stored at individual nodes, and to minimize the amount of data passed in the network. Therefore, the amount of energy required by individual nodes for data transmission is reduced. Furthermore, with the dynamic selection of the nodes needed for classification, the overall number of active nodes is reduced. This would lead to reducing the overall power consumption of the system and can potentially increase system lifetimes.

6.2 HUMAN ACTIVITY MONITORING

Human action recognition can employ data from either environmental or on-body sensing devices. The fields of computer vision and surveillance have traditionally used cameras as environmental sensors to monitor human movement [10]. Processing image sequences to recognize motions is the foundation of this approach. In tracking systems, cameras are used to detect actions performed by subjects inside the observed area. The use of video streams in wireless networks to interpret human motion has been considered for assisted-living applications [11, 12]. In contrast, BSNs are built of lightweight sensor platforms, which aim to recognize the actions performed by the person wearing them [13]. The sensors can be mounted on the human body or clothing or even woven into the clothing itself [14]. Unlike vision-based platforms, BSNs require no environmental infrastructures. They are also less expensive. Moreover, the signal readings from on-body sensors are unbiased by environmental effects such as light and the background. This makes BSNs potentially more feasible than vision frameworks for continuous and remote health-care monitoring.

A number of researchers have integrated on-body sensors in a wireless network for the purpose of activity recognition and lifestyle monitoring. Logan et al. [15] reported the results of a study on activity recognition using different types of sensory devices, including built-in wired sensors, RFID tags, and wireless inertial sensors. The analysis performed on 104 hours of data collected from more than 900 sensor inputs shows that motion sensors outperform the other sensors on many of the movements studied. A prototype called *MEDIC*, developed by Wu et al. [16] for remote health-care monitoring, uses a PDA (personal digital assistant) as the base station and several sensor nodes that collect and process physiological data. They use a naive Bayes classifier [17] that provides more than 90% accuracy. A wireless body sensor system for monitoring human activities and location in indoor environments was introduced by Klingbeill and Wark [18], where each sensor node is equipped with accelerometer, gyroscope, and magnetometer. Bao and Intille [19] used a network of five accelerometers to classify a sequence of daily activities. They report a classification accuracy of 84% for detecting 20 actions. The system described by Lester, Choudhury, and Borriello [20] uses seven different sensors embedded in a single node to classify 12 movements. The accuracy obtained by this system is 90%.

Most methodologies for representing human movements map all sensor readings to an identical feature space and then use traditional classification algorithms such as *k*-nearest-neighbor and naive Bayes classifiers [21] to detect movements [16, 22]. While these statistical models provide acceptable classification accuracies, they suffer from complexities involved in representing actions in terms of a high-dimensional feature space. Even applying feature reduction techniques such as principal component analysis (PCA) [23] leaves each node to transmit the features to a base station for the purpose of classification. A more efficient form of representing movements is a linguistic framework where signal readings are transformed into a set of primitives. Motivation for such an approach comes from the fact that movement and spoken language use a similar cognitive substrate in terms of grammatical hierarchy [24]. The task is called *phonology*, which finds basic primitives, similar to phonemes, and assigns appropriate symbols to them.

6.3 WEARABLE COMPUTERS AND DISTRIBUTED CLASSIFICATION

Action recognition based on wearable computers uses a BSN consisting of several sensor units in a wireless network. The sensor nodes aim to collectively detect transitional movements according to a training model. Each node, which is also called a *mote*, has one or more motion sensors, for example a triaxial accelerometer, a biaxial gyroscope, a microcontroller, and a radio, as shown in Figure 6.1. Nodes sample sensor readings at a certain frequency and can transmit the data wirelessly to each other. One may design and manufacture his/her own sensor board from scratch rather than using off-the-shelf sensor nodes. An example of a mote is TelosB [25], which is commercially available from XBow®. The core processing unit of a Telos mote is TI MSP430, with active power of 3 mW. The power consumption of the radio is 38 mW and 35 mW in receive and transmit modes, respectively. Each mote has a custom-designed sensor board and is powered by a Li-ion battery.

A decision-based distributed classification algorithm uses the signal-processing model shown in Figure 6.2 to classify an unknown action by integrating information from sensors across the network. This processing model requires several parameters

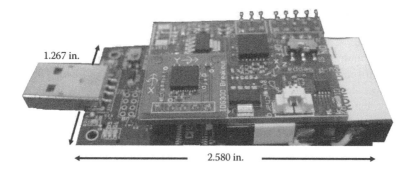

FIGURE 6.1 A sensor node composed of a processing unit and a custom-designed sensor board. The motion sensor board has a triaxial accelerometer and a biaxial gyroscope.

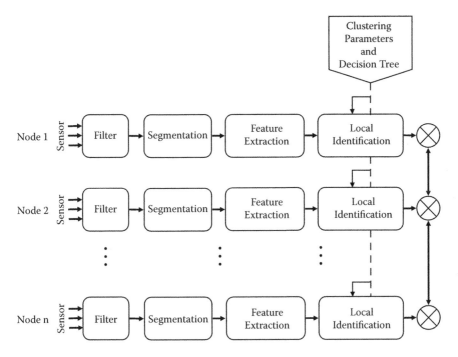

FIGURE 6.2   Signal-processing flow for distributed action recognition based on decision-tree model.

as well as a decision tree, which are precalculated during training. The training is conducted as shown in Figure 6.3. A brief description of each processing task is given in the following.

Sensor data collection: Data are collected from each of the sensors (e.g., x, y, z accelerometer and x, y gyroscope) on each of the sensor nodes at a certain sampling rate. Studies have shown that a sampling frequency of 50 Hz is sufficiently large to provide high resolution of human motion data while compensating for the bandwidth constraints of the sensor platform. It further satisfies the Nyquist criterion [26].

Preprocessing: The data are filtered to remove high-frequency noise. Studies suggest a five-point moving-average filter for this purpose [27]. The number of points used to average the signal is usually chosen by examining the power spectral density of the signals. The filter is required to remove unnecessary artifacts (such as wrist tremors) while maintaining significant data.

Segmentation: This processing task aims to determine the temporal region of the signal that represents an action. Segmentation is not the focus of this chapter. Much research has been done on developing effective segmentation approaches, such as *fixed size segments* [19], *segmentation based on signal energy* [28, 29], and *segmentation based on Hidden Markov Models* [30, 31].

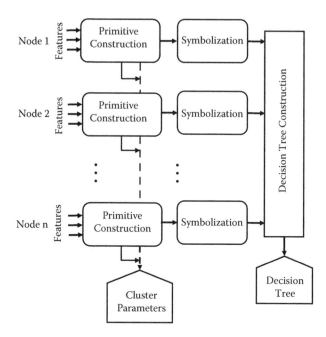

FIGURE 6.3   Training model to construct action primitives and build a decision tree.

Feature extraction: Single-value features are extracted from the signal segment. Statistical features include mean, start-to-end amplitude, standard
deviation, peak-to-peak amplitude, and root-mean-square (RMS) power.
Moreover, morphological features can be obtained from uniformly distributed points on the signal, where each feature is the value of the signal at
one of the points. Intuition behind using this feature set is that (1) these
time-domain features are computationally inexpensive and can be executed
on the lightweight sensor nodes, (2) they capture both statistical and structural properties of the signal, and (3) the significance of statistical and morphological features for human movement analysis has been suggested by
researchers [27, 32].

Per-node identification: Each node uses a phonetic expression of movements
to map a given action onto the corresponding primitive. Each primitive is
represented by a cluster, which groups similar actions together based on the
features that are extracted from each signal segment. During identification,
an unknown action is mapped onto a cluster based on the cluster configuration obtained during training.

Distributed action recognition: The global state of the system is recognized
by combining local knowledge from different nodes using a decision-tree
model. The decision tree allows for incremental classification of actions
and is specifically constructed to minimize the number of nodes involved
in the classification. Each node can distinguish certain actions based on the
cluster assignment of the actions. The node that is most informative among

all existing nodes is used as the root of the tree. The remaining nodes are organized in the tree according to their capabilities in distinguishing the rest of the actions. This decision-tree-based model classifies an unknown action by extracting information from a subset of nodes in a certain order and typically determining the action before all nodes are considered. The decision tree, which is constructed during training, determines the node that initiates communication. The algorithm proceeds by transmitting local results of one node to the next node in the tree. Upon receiving data, the node combines the data with its local statistics and may decide to branch to another node in the tree. This process continues until all the actions are distinguished, detecting the target action.

Primitive construction and symbolization: A phonological approach is used to construct spatially distributed primitives of actions at each sensor node. Primitives are created using unsupervised classification techniques that map signal readings with similar patterns onto the same clusters. For example, a *K*-means clustering algorithm can be used to group similar actions based on the value of the statistical and morphological features extracted from each of the signal segments. On each individual sensor node, several actions might fall into the same cluster. Each cluster represents a primitive. When primitives are generated, a symbol is assigned to each primitive. By symbolization, each action is represented in terms of a set of meaningful phonemes that are labeled by alphabets.

Accurate detection of actions requires a global view of the whole system, but each individual node in a BSN has only local knowledge of the event taking place. This makes the problem of training and recognition very challenging [33]. The amount of knowledge provided by each node determines the usefulness of that node in recognizing actions. It seems that the right place to begin a discussion on phoneme construction for distributed wearable computers is to consider the properties of movements and the relationship between each movement and the system configuration.

Each node in the system has a different perspective on the action. The ability of a node to recognize actions varies based on the type of action. For example, consider the two actions "stand to sit" and "bend." A node mounted on the arm might distinguish these two actions, but a sensor on the ankle might not provide useful information. The main advantage of the clustering model is that by grouping training trials at each node, nodes not contributing to a certain action will be identified. That is, one will be able to determine which nodes are useful for recognizing which actions.

6.4 PHONOLOGICAL APPROACH TO MOVEMENT REPRESENTATION

The idea of primitive-based activity recognition originated from research work in computer vision, where both static and dynamic vision-based approaches have been developed. For static methods, individual time frames of a video sequence are used as the basic components for analysis. Recognition involves the combination of discrete information extracted from individual frames. In dynamic methods, a fixed interval of a video stream is the major unit of analysis. The Hidden Markov Model (HMM) [34], which takes into account the correlation between adjacent time instances by formulating a Markov process [35], is often used for the dynamic representation of motion due to its ability to handle uncertainty with its stochastic framework.

Examples include the work presented by Niwase, Yamagishi, and Kobayashi [36] that introduces a statistical technique for synthesizing walking patterns. The motion is expressed by a sequence of primitives extracted using an HMM-based model. The limitation of HMM in efficiently handling several interdependent processes has led to the creation of grammar-based representations of human actions. Guerra-Filho, Fermuller, and Aloimonos [37] proposed a platform for a visuo-motor language. They provide the basic kernel for the symbolic manipulation of visual and motor information in a sensory-motor system. Their phonological rules are modeled as a finite automaton. Reng, Moeslund, and Granum [38] present an algorithm that finds primitives for human gestures using a motion-capture system that measures the three-dimensional (3-D) position of body parts. The trajectory of motion of these parts is considered as a gesture, and primitives are constructed based on the density of the training data set.

These vision-based techniques tend to discover grammatical properties within human activities. In BSNs, however, deployment of visual sensors will both restrict the system to less mobility and affect its wearability. Furthermore, in contrast to vision-based approaches that construct action primitives from a time-series video, sensor nodes are spatially distributed in a BSN. Therefore, spatial distribution of the nodes must be taken into consideration when constructing primitives in BSN platforms.

Although the aforementioned linguistic methods have been successful in providing a grammatical platform for action recognition, they could not be efficiently employed in systems based on BSNs because they are explicitly designed for visual data. Moreover, action recognition using BSNs requires efficient techniques for information fusion of distributed sensor nodes. Since these platforms are energy-constrained and have limited computational power, distributed algorithms that activate only a subset of nodes for classification and need less communication are desirable.

6.4.1 CLUSTERING IMPLICATIONS

Clustering algorithms are employed to find primitives for each action. This section describes how the system benefits from clustering algorithms and strategies of finding the most effective clustering configuration.

6.4.1.1 Clustering Techniques

Clustering is grouping together data points in a data set that are most similar to each other. Two major clustering techniques are hierarchical clustering [39] and K-means clustering [40]. In the hierarchical method, each data item is initially considered a single cluster. At each stage of the algorithm, similar clusters are grouped together to form new clusters. In the K-means algorithm, K centroids are chosen; one for each cluster. In this way, training data are grouped into a predefined number of clusters. Unlike hierarchical clustering, in which clusters are not improved after being created, the K-means algorithm iteratively improves the initial clusters. The continuously improving nature of this algorithm leads to high-quality clusters when provided appropriate data. In what follows, the K-means clustering algorithm is used for primitive construction, mainly because it is simple and operates based on the firm foundation of analysis of variances [41].

6.4.1.2 Cluster Validation

Although K-means is a popular clustering technique, the partition attained by this algorithm is dependent on both the initial value of centroids and the number of clusters. To increase the likelihood of arriving at a good partitioning of the data, many improvements to K-means have been proposed in the literature. The sum of square error (SSE) is a reasonable metric used to find the global optimal solution [42]. To cope with the effects of initialization, one approach is to use uniformly distributed initial centers [43] and repeatedly search for the configuration that gives the minimum error. The SSE error function is calculated as in Equation 6.1, where x_i denotes the ith data item, μ_k denotes the centroid vector associated with cluster C_k, and K is the total number of clusters.

$$\text{SSE} = \sum_{k=1}^{K} \sum_{i \in C_k} (x_i - \mu_k)^2 \tag{6.1}$$

Another problem with K-means is that it requires prediction of the correct number of clusters. Usually, a cluster-validity framework provides insight into this problem. The silhouette quality measure [44] is an effective measure for this purpose. It is robust and takes into account both intracluster and intercluster similarities to determine the quality of a cluster. Using nonnormalized features, this metric is calculated in Euclidean space [45]. Let C_k be a cluster constructed by the K-means algorithm. The silhouette measure assigns a quality metric S_i to the ith data item of C_k. This value signifies the confidence of the membership of the ith item to cluster C_k. S_i is defined by Equation 6.2, where a_i is the average distance between the ith data item and all of the items inside cluster C_k, and b_i is the minimum of the average distances between the ith item and all of the items in each cluster besides C_k. That is, the silhouette measure compares the distance between an item and the other items in its assigned cluster to the distance between that item and the items in the nearest neighboring cluster. The larger the S_i, the higher the level of confidence about the membership of the ith sample in the training set to cluster C_k.

$$S_i = \frac{b_i - a_i}{\max\{a_i, b_i\}} \tag{6.2}$$

While S_i, also called *silhouette width*, describes the quality of the membership of a single data item, the quality measure of a partition, called the *silhouette index*, for a given number of clusters K is calculated using Equation 6.3, where N is the number of data items in the training set.

$$\text{Sil}(K) = \frac{1}{N} \sum_{i=1}^{N} S_i$$ (6.3)

To obtain the most effective configuration in terms of the number of clusters, one can choose the K that has the largest silhouette index, as shown in Equation 6.4.

$$\hat{K} = \arg\max_{K}\{\text{Sil}(K)\}$$ (6.4)

6.4.2 PHONOLOGY

Phonology refers to identifying basic primitives of actions. In a wearable networked framework, primitives are distributed among sensor nodes with varying ability to recognize actions. A phonetic description is able to characterize each action in terms of primitives using a two-stage process that is described in the following sections.

6.4.2.1 Primitive Construction

Primitives are created by individual sensor nodes during training. K-means is used to perform local clustering at each node, transforming the feature space into groups of dense data items. Each cluster is associated with a primitive. This technique is effective, since it provides insights into the usefulness of nodes for detecting each action. Actions with similar patterns tend to be assigned to the same cluster at each node, while they might be represented by different clusters at another node. Let $A = \{a_1, a_2,..., a_m\}$ be a set of m actions to be classified. The clustering algorithm at node s_i will transform the actions into a series of clusters $\{P_{i1}, P_{i2},..., P_{ic}\}$, where the number of clusters c is limited to be at most m, the number of actions. The intuition behind this constraint is that similar actions will be grouped and, therefore, the total number of clusters will be less than the number of actions. The validation techniques explained in Section 6.4.1.2 are then used to find the most effective clustering configuration.

6.4.2.2 Symbolization

The second step in constructing a phonetic description is to select a final group of primitives and assign symbols to them. Some of the clusters defining the initial primitives are of low quality, meaning that the primitives they define will not be good representations of the human actions. The clusters, however, can be refined by calculating the silhouette quality measure for each cluster and eliminating clusters that do not meet a certain threshold. The threshold is usually chosen to guarantee that each action falls into at least one cluster [27]. In this way, the set of primitives at node i might be reduced to $\{P_{i1}, P_{i2},..., P_{ip}\}$, where $p \leq c$ is the number of final primitives after applying the quality measure, and c is the number of original primitives. After cluster refinement, each cluster P_{ir} ($r \in \{1,2,...,p\}$) is assigned a unique symbol ρ_{ir} from an alphabet Σ [46].

TABLE 6.1

Notations

Symbol	Description
$S = \{s_1, s_2, \ldots, s_n\}$	set of n sensor nodes that form a BSN
i and j	indices used for sensor nodes (e.g., s_i, s_j)
n	total number of sensor nodes
$A = \{a_1, a_2, \ldots, a_m\}$	set of m actions/movements to be detected
k and l	indices used for actions (e.g., a_k, a_l)
m	total number of actions
$P = \{\rho_{ir}\}$	set of primitives created by sensor node s_i
r and t	indices used for primitives (e.g., ρ_{ir}, ρ_{it})

6.5 DECISION-TREE CONSTRUCTION

Physical movement monitoring by sensor networks requires the combination of local knowledge from each node to achieve a global view of human behaviors. This section studies the problem of constructing a decision tree for action recognition based on the semantic subspace generated by the primitives. The notations in Table 6.1 are used throughout this section.

6.5.1 DECISION PATH FOR ACTION RECOGNITION

The problem of recognizing actions using primitives can be viewed as a decision-tree problem in which each internal decision node represents a sensor node and its branches are the primitives identified within that node. The terminal leaves correspond to the actions to be identified. It is required that the tree identifies each action correctly. The aim of action recognition is to assign an unknown action to one of m mutually exclusive actions. The ordering of nodes in the tree changes its height and thus the time needed to converge to a solution. Given a decision tree T, path length for an action a_k ($a_k \in A$; A denotes set of all actions, as shown in Table 6.1) is defined to be the number of internal nodes in the path from root to a_k. The path length for action a_k is denoted by $\ell(a_k)$. The cost of the tree is the sum of all path lengths.

$$\text{Cost}(T) = \sum_{a_k \in A} \ell(a_k) \qquad (6.5)$$

Before constructing a full decision tree, it would be nice to first seek a linear ordering of the nodes that minimizes convergence time. Using this model, the recognition policy will explore a predefined series of sensor nodes regardless of observations made by individual nodes in real time. Therefore, such classification is static in terms of the ordering. An optimal ordering of the sensor nodes is then a minimum-cost decision path. In an effort to make this model more efficient, a full decision tree

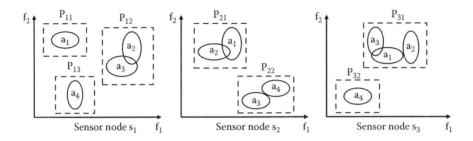

FIGURE 6.4 An example of three nodes (s_1, s_2, s_3) and four actions (a_1, a_2, a_3, a_4). The actions are mapped to seven primitives $(\rho_{11}, \rho_{12}, \rho_{13}, \rho_{21}, \rho_{22}, \rho_{31}, \rho_{32})$. Each action is symbolized by corresponding primitives; $a_1 = \{\rho_{11}, \rho_{21}, \rho_{31}\}$; $a_2 = \{\rho_{12}, \rho_{21}, \rho_{31}\}$; $a_3 = \{\rho_{12}, \rho_{22}, \rho_{31}\}$; $a_4 = \{\rho_{13}, \rho_{22}, \rho_{32}\}$.

needs to be constructed in which different branches can be taken according to the accumulative knowledge attained by currently visited sensor nodes.

To better illustrate the identification problem, a simple example, shown in Figure 6.4, is discussed here. The system consists of three sensor nodes denoted by s_1, s_2, and s_3 and four actions denoted by a_1, a_2, a_3, and a_4, which are shown using a two-dimensional feature space (f_1 and f_2). The ellipsoids depict the distribution of different classes across the network. Dashed-line rectangles show the mapping of actions to primitives. The system has seven final primitives denoted by $\{\rho_{11}, \rho_{12}, \rho_{13}, \rho_{21}, \rho_{22}, \rho_{31}, \rho_{32}\}$. In node s_1, action a_1 is mapped to primitive ρ_{11}, actions a_2 and a_3 are mapped to primitive ρ_{12}, and action a_4 is mapped to primitive ρ_{13}. In other nodes, the actions are mapped to primitives as shown. This phonetic expression can effectively describe the ability of individual nodes to identify the actions. For instance, node s_1 can distinguish action a_1 from the rest of the actions, as it finds no ambiguity when mapping an action to ρ_{11}, but it cannot distinguish between actions a_2 and a_3, as they are mapped to the same primitive. While each node has limited knowledge of the system, we require a global view in which every action is distinguished from the rest. Furthermore, an ordering of sensor nodes that minimizes the total time of convergence is required.

Given an instance of a decision problem, one can construct different decision trees. Figure 6.5 illustrates a sample decision tree for the example represented in Figure 6.4. The problem of finding a minimal decision tree is shown to be hard to approximate [47]. Therefore, this section investigates construction of a decision path for action recognition. The method of linearly ordering the nodes restricts the shape of the decision tree so that all nodes are placed on a single path from the root, and the tree has a height equal to the total number of nodes required for recognition. The construction of a full decision tree is discussed in Section 6.5.5.

6.5.2 PROBLEM FORMULATION

A formal definition of the action identification problem using a decision path is presented in this section.

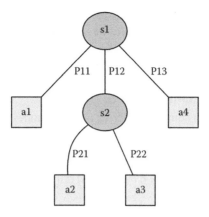

FIGURE 6.5 A sample decision tree for the example illustrated in Figure 6.4.

6.5.2.1 Definition 1: (Local Discrimination Set)

Let $A = \{a_1, a_2, \ldots, a_m\}$ be a finite set of actions mapped to a set of primitives $P = \{\rho_{ir}\}$. The local discrimination set LDS_i for a node s_i is defined by

$$LDS_i = \{(a_k, a_l) \mid a_k \in \rho_{ir}, \quad a_l \in \rho_{it}, \quad r \neq t, \quad k < l\}. \tag{6.6}$$

The LDS_i expresses the pairs of actions that can be distinguished by s_i. In the example shown in Figure 6.4, the local discrimination set for node s_1 is $LDS_1 = \{(a_1,a_2), (a_1,a_3), (a_1,a_4), (a_2,a_4), (a_3,a_4)\}$. For the other nodes, we have $LDS_2 = \{(a_1,a_3), (a_1,a_4), (a_2,a_3), (a_2,a_4)\}$, and $LDS_3 = \{(a_1,a_4), (a_2,a_4), (a_3,a_4)\}$.

6.5.2.2 Definition 2: (Global Discrimination Set)

Let $A = \{a_1, a_2, \ldots, a_m\}$ be a finite set of actions and $P = \{\rho_{ir}\}$ a collection of primitives. The global discrimination set GDS is defined by

$$GDS = \{(a_k, a_l) \mid a_k \in A, a_l \in A, \ k < l\}. \tag{6.7}$$

The global discrimination set contains all the pairs of actions that are required to be distinguished from one another. For example, the global discrimination set for the system shown in Figure 6.4 is $GDS = \{(a_1,a_2), (a_1,a_3), (a_1,a_4), (a_2,a_3), (a_2,a_4), (a_3,a_4)\}$. In this example, the objective is to distinguish between every pair of actions.

6.5.2.3 Definition 3: (Complete Ordering)

Let $A = \{a_1, a_2, \ldots, a_m\}$ be a finite set of actions and $P = \{\rho_{ir}\}$ a collection of primitives. An ordering $O = \{s_1, s_2, \ldots, s_n\}$ is complete if the following condition holds.

$$\bigcup_{i=1}^{n} LDS_i = GDS. \tag{6.8}$$

This indicates that the ordering is capable of distinguishing between all required pairs of actions. In the previous example, the ordering $O = \{s_1, s_2\}$ is complete, since $LDS_1 \cup LDS_2 = GDS$, but the ordering $O = \{s_2, s_3\}$ is not complete because this ordering cannot discriminate between actions a_1 and a_2.

6.5.2.4   Definition 4: (Ordering Cost)

Let $O = \{s_1, s_2,..., s_n\}$ be a complete ordering of sensor nodes and $f(a_k)$ a function that gives the index of the first node in which the following condition holds:

$$\{(a_k, a_l) \mid k < l, \quad a_k \in A\} \subset \bigcup_{i=1}^{f(a_k)} LDS_i. \tag{6.9}$$

That is, $f(a_k)$ is the number of nodes required to distinguish a_k from all other actions. Then the total cost of the ordering is given by the following equation:

$$Z = \sum_{a_k \in A} f(a_k). \tag{6.10}$$

This formulation weights the cost of an ordering so that an ordering in which more actions require fewer nodes has a lower cost. For instance, let $O = \{s_3, s_2, s_1\}$ be a complete ordering for the example shown in Figure 6.4. Then $f(a_4) = 1$ because action a_4 can be completely identified by the first visited node (s_3). At the next node (s_2), action a_3 can be distinguished from the remaining actions (a_1 and a_2). Thus, $f(a_3) = 2$ because action a_3 is identified at the second node. Finally, actions a_1 and a_2 will be detected by visiting the third node (s_1), meaning that $f(a_1) = f(a_2) = 3$. Therefore, the total cost for this ordering is 9.

6.5.2.5   Definition 5: (Min Cost Identification Problem)

Given a finite set GDS and LDS, where $LDS = \{LDS_1, LDS_2,..., LDS_n\}$ is a collection of subsets of GDS such that the union of all LDS_i forms GDS, Min Cost Identification (MCI) is the problem of finding a complete linear ordering such that the cost of the ordering is minimized.

In the previous example, it would be easy to find the optimal solution by a brute-force technique. We can see that the cost for the optimal ordering (s_1, s_2) is 6.

6.5.3   Problem Complexity

In this section, the complexity of Min Cost Identification is discussed. It will be shown that this problem is NP-hard (nondeterministic polynomial-time hard) by reduction from Min Sum Set Cover.

6.5.3.1   Definition 6: (Min Sum Set Cover)

Let U be a finite set of elements and $S = \{S_1, S_2,..., S_m\}$ a collection of subsets of U such that their union forms U. A linear ordering of S is a bijection f from S to $\{1, 2,..., m\}$.

For each element $e \in U$ and linear ordering f, let us define $f(e)$ as the minimum of $f(S)$ over all $\{S_i : e \in S_i\}$. The goal is to find a linear ordering that minimizes $\Sigma_e f(e)$.

Theorem 1. *The Min Cost Identification problem is NP-hard.*

Proof. We prove that the Min Cost Identification problem is NP-hard by reduction from Min Sum Set Cover (MSSC). Consider an MSSC instance (U, S) consisting of a finite set of elements U and a collection S of subsets of U. The objective is to find a minimum-cost linear ordering of subsets such that the union of the chosen subsets of U contains all elements in U. Let us now define a set $\tilde{U}$ by replacing elements of U with all elements (a_k, a_l) from the GDS. Let us also define $\tilde{S}$ by replacing its subsets S_i with LDS_i. $(\tilde{U}, \tilde{S})$ is an instance of the MCI problem. Therefore, MCI is NP-hard. Since solutions for the decision problem of MCI are verifiable in polynomial time, it is in NP, and consequently, the MCI decision problem is also NP-Complete. ∎

Theorem 2. *There exists no polynomial-time approximation algorithm for MCI with an approximation ratio less than* 4.

Proof. The reduction from MSSC to MCI in the proof of Theorem 1 is approximation preserving; that is, it implies that any lower bound for MSSC also holds for MCI. In the literature [48], it is shown that for every $\varepsilon > 0$, it is NP-hard to approximate MSSC within a ratio of $4 - \varepsilon$. Therefore, 4 is also a lower bound for the approximation ratio of MCI. ∎

6.5.4 GREEDY SOLUTION

The greedy algorithm for MCI is adapted from the greedy algorithm for MSSC and is shown in Algorithm 1. At each step, it searches for the node that can distinguish between the maximum number of remaining actions. It then adds such a node to the solution space and removes the actions it distinguishes from further consideration. The algorithm terminates when all pairs of actions are distinguished from each other. The approximation ratio is 4, as previously discussed.

ALGORITHM 1
Greedy solution for MCI

```
Require: Set of actions A, set of primitives P, and set of
nodes S
Ensure: Linear ordering O
        calculate set LDS_i for every node s_i
        calculate set GDS
        O = φ
        while O ≠ GDS do
```

```
                take node s_i such that LDS_i is maximum cardinality
        O = O ∪ s_i
        for all e ∈ LDS_i do
                remove e from all LDS_j (j = {1,…, n})
        end for
    end while
```

Algorithm 1 can be used to find the minimum number and preferred locations of sensor nodes required to recognize certain actions. This can be used for power optimization because, at any time, only a subset of sensor nodes will be required to be active, based on the actions of interest at that time. Furthermore, reducing the number of required nodes helps in enhancing the wearability of the BSN platform.

6.5.5 FULL DECISION TREE FOR ACTION RECOGNITION

The linear ordering of the sensor nodes provides a decision tree that explores a predefined series of the nodes. When looking for the next node to visit, the algorithm that constructs the decision tree does not take into account the primitives to which an action might be mapped. The convergence time can be further reduced if information on primitives is taken into consideration. Each primitive within a sensor node is a mapping of several actions. An unknown action is detected by following a path in the tree starting from the root and ending at a leaf node. An action mapped onto a certain primitive within may require a different path than an action mapped to another primitive. Therefore, constructing a minimum-cost decision tree is required to guarantee the fastest possible action recognition.

Obtaining an optimal decision tree has been shown to be NP-hard. In the following discussion, we present the results of hardness approximation for this problem introduced in the literature [49, 50]. Chakaravarthy et al. [50] have studied a decision-tree problem for entity identification where an input table represents m attributes (columns) of N entities (rows). To identify an unknown entity, a decision tree is required in which each internal node is labeled with an attribute, and its branches are labeled with the values that the attribute can take. The entities are placed in the leaves of the tree. The cost of a decision tree is the expected distance of an entity from the root. The goal is to construct a minimum-cost decision tree. The case of the problem where entities are equally likely is called *UDT* [50].

Theorem 3. *For any $\varepsilon > 0$, it is NP-hard to approximate the UDT problem within a ratio of $(4 − \varepsilon)$ [50].*

The problem of constructing a minimum-cost decision tree for action recognition is similar in spirit to the UDT problem stated previously. Sensor nodes, primitives, and actions in the primitive identification problem correspond to attributes, values, and entities in the UDT problem. Therefore, the same approximation ratio holds for the full decision-tree construction problem.

Theorem 4. *For any ε > 0, it is NP-hard to approximate the problem of constructing a full decision tree for action recognition within a ratio of (4 − ε).*

A greedy algorithm is presented here to construct a minimum-cost decision tree for the recognition problem. The algorithm is adapted from Chakaravarthy et al. [50] for entity identification and is shown in Algorithm 2.

ALGORITHM 2
Greedy approximation for full decision tree

```
Require: α ⊂ A, a subset of actions A = {a₁, a₂,…, aₘ} to be identified
Ensure: Decision tree T
        if |α| = 1 then
              T is a single node aₖ ∈ A
        else
              let sᵢ be the sensor node whose LDSᵢ is maximum cardinality
              create root node sᵢ

              for all e ∈ LDSᵢ do
                    remove e from all LDSⱼ (j = {1,…, n})
              end for

              for all r ∈{1,…, B} do
                    let αᵢᵣ = {aₖ| aₖ ∈ ρᵢᵣ}
                    Tⱼ = Greedy(αᵢᵣ)
                    let sⱼ be the root of Tⱼ
                    add Tⱼ to T by adding a branch from sᵢ to sⱼ
              end for
        end if
```

6.5.5.1 Definition 7: (Branching Factor)

Let $A = \{a_1, a_2,\ldots, a_m\}$ be a set of actions, and let $\Pi_i = \{\rho_{i1}, \rho_{i2},\ldots, \rho_{ip}\}$ denote a set of primitives associated with sensor node s_i. Then the branching is the maximum number of distinct primitives among all sensor nodes, and is given by Equation (6.11).

$$B = \arg\max_i |\Pi_i| \tag{6.11}$$

The intuition behind the greedy approximation is that a good decision tree should distinguish pairs of actions at higher levels of the tree. Therefore, a natural idea is to make the sensor node that distinguishes the maximum number of pairs as the root of the tree. Assigning a sensor node s_i as the root node will partition the set of actions A into disjoint sets $\alpha_{i1}, \alpha_{i2},\ldots, \alpha_{iB}$, where $\alpha_{ir} = \{a_k \mid a_k \in \rho_{ir}\}$. The same greedy approximation is applied to each of these sets to obtain B decision trees and make them the subtrees of the root node.

Theorem 5. *The greedy algorithm has an approximation ratio of O(rB log m)* (see Chakaravarthy et al. [50] for proof).

In the approximation ratio for Algorithm 2, B is the branching factor, m denotes the number of input actions, and rB is a Ramsey number [51] that can be suitably defined and has the value of at most log B.

6.5.6 Distributed Classification Algorithm

A distributed algorithm is presented in this section. The algorithm uses the decision-tree model for action recognition. It assumes that each node processes data locally and maps an unknown action to the clusters generated during training. It further assumes that all the nodes are perfectly synchronized prior to execution of the distributed algorithm. This can be done during segmentation to ensure that all the nodes are collaborating for classification of the same action. Communication is initiated by the most informative node, which is located at the root of the tree. The root node generates a token that is transmitted across the network as the classification algorithm proceeds. The computation is executed by a series of the nodes until the solution converges. Each sensor node maintains a data structure, including the decision-tree structure, its local computation, and statistics received from other nodes. In particular, each node s_i keeps track of recognition convergence by a variable Target Movement Vector (TMV), which initially contains all actions as possible target movements. As the algorithm proceeds, each node may decide to discard some actions from the TMV according to the cluster membership of the action. The algorithm takes three steps as follows:

Initialization: Each sensor node s_i assigns an unknown action a_k to a cluster indicated by primitive ρ_{ir}. It further updates TMV by rejecting all actions a_l that do not belong to the cluster ρ_{ir}. Moreover, the root node generates a token that enables internode communication.

Transmission: A sensor node s_i transmits its local statistics only if it is the current owner of the token t. The data, including updated TMV$_i$ and the token, are then transmitted to the next node in the decision tree. The next node is determined according to the assignment of the current assignment of the unknown action (ρ_{ir}).

Update: Upon receiving data, each node s_j updates its local Target Movement Vector TMV$_j$ by combining the results provided by the sender node s_i. That is, s_j might reject further actions from consideration in subsequent steps of the algorithm. The receiver also checks conditions for termination. Specifically, it checks the convergence vector TMV$_k$, which contains the possible actions left. If only one action is left in the vector, the node declares a convergence and reports that action as the target action. However, if more than one action is left in TMV$_k$, the node would find the next node in the tree for data transmission.

6.6　PERFORMANCE OF DECISION-TREE-BASED CLASSIFICATION

This section describes experimental procedure and results that demonstrate the effectiveness of the developed action-recognition algorithms.

The performance results presented in this section are based on real data collected from human subjects while using wearable motion sensors and performing transitional movements. These experiments were conducted using a BSN composed on the sensor nodes described in Section 6.3. Five subjects aged between 22 and 55 wore nine sensor nodes, as shown in Figure 6.6. The subjects performed the actions listed

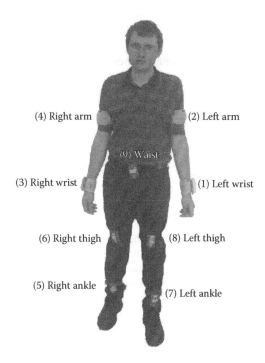

(4) Right arm   (2) Left arm

(9) Waist

(3) Right wrist   (1) Left wrist

(6) Right thigh   (8) Left thigh

(5) Right ankle
                                                   (7) Left ankle

FIGURE 6.6   Experimental subject wearing nine sensor nodes.

in Table 6.2 for 10 trials each. The experiments consisted of a relatively wide range of actions that required motions from different segments of the body.

The motes were programmed to sample sensors (accelerometer and gyroscope) at 50 Hz. The sampling frequency was chosen to satisfy the Nyquist criterion. The sampled data were sent wirelessly to a base station using a TDMA (time-division multiple access) protocol. The base station is connected to a laptop via USB to deliver received data to the data-collector tool that splits the collected data into different files according to the type of movement and the sensor node that has transmitted the data.

It is quite common to use 50% of the data collated from all subjects as a training set and 50% as a test set. The training set is used for constructing primitives and finding the minimum-cost ordering of the nodes as well as a full decision tree, while the test set is used to verify the accuracy of the recognition technique. For each trial, the raw sensor readings are passed through a moving-average filter to reduce high-frequency noise.

6.6.1 CONSTRUCTED PRIMITIVES

As previously stated, *K*-means clustering is used at each sensor node to create action primitives. The clustering-refinement approach would map each action to one of the generated clusters. Figure 6.7 shows how different actions are mapped to the

TABLE 6.2

Transitional Actions Performed for Experimental Analysis

Action No.	Description
1	Stand to sit
2	Sit to stand
3	Stand to sit to stand
4	Sit to lie
5	Lie to sit
6	Sit to lie to sit
7	Bend and grasp
8	Kneeling, right leg first
9	Kneeling, left leg first
10	Turn clockwise 90°
11	Turn counterclockwise 90°
12	Turn clockwise 360°
13	Turn counterclockwise 360°
14	Look back clockwise
15	Move forward (1 step)
16	Move backward (1 step)
17	Move to the left (1 step)
18	Move to the right (1 step)
19	Reach up with one hand
20	Reach up with two hands
21	Grasp an object with right hand, turn 90°, and release
22	Grasp an object with two hands, turn 90°, and release
23	Jumping
24	Going upstairs (2 stairs)
25	Going downstairs (2 stairs)

constructed primitives. A black cell represents mapping of an action to the corresponding primitive (ρ_{ij}). The number of primitives ranges from five for Left-thigh and Right-thigh to eight for Left-wrist, Right-arm, and Left-ankle. The number of primitives per node and the actions that are mapped to each cluster depend on the ability of the node in distinguishing between different actions. We note that actions a_{24} and a_{25} (going upstairs and downstairs) are mapped to the same cluster on every individual node except the Right-ankle node, which means that the Right-ankle is the only node that can distinguish these two actions and, therefore, is needed for detection of a_{24} and a_{25}. Also, the four actions a_{15} to a_{18} (moving forward, backward, or to the side) are associated with the same cluster on all the nodes except Right-ankle and Left-ankle nodes, meaning that the ankle nodes contribute most to detection of these actions.

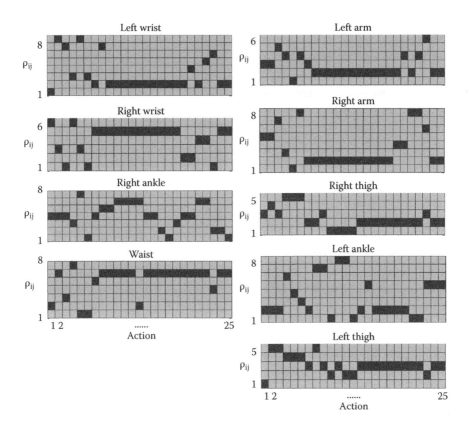

FIGURE 6.7   Primitives and corresponding mapping of the actions for each sensor node.

6.6.2 CONSTRUCTED DECISION TREES

The greedy solution described in Algorithm 1 is used to solve the Min Cost Identification (MCI) problem. Solution to this problem is a linear ordering of the nodes that can be used for action recognition. Given that the system consisted of nine sensor nodes to recognize 25 actions, the size of the GDS set is 300, which is the total number of action pairs (a_k, a_l) to be distinguished. The ordering obtained by the greedy algorithm is given in Table 6.3. It shows that in the worst case, six sensor nodes were sufficient to achieve a global knowledge of the current event in the system. The most informative node is the node placed on the Left-ankle with ordering 1, meaning that it needs to be the first node to visit. The value of the cost function associated with this node is 3 (last column in Table 6.3) because it could alone distinguish three actions (actions 4, 5, and 6, as shown in Figure 6.8) from all others actions. The second node on the decision path is the Right-wrist node, which can make a classification decision about four other actions (actions 1, 2, 3, and 23). The value of the cost function for this node is 8, which together with the first node

TABLE 6.3

Statistics for Greedy Algorithm

Node#	Position	#Primitives	Ordering	#Actions	$\sum_{a_k \in A} f(a_k)$
1	Left wrist	8	...	...	...
2	Left arm	6	3	5	26
3	Right wrist	6	2	4	11
4	Right arm	8	...	...	...
5	Right ankle	7	4	7	54
6	Right thigh	5	...	...	...
7	Left ankle	8	1	3	3
8	Left thigh	5	5	4	74
9	Waist	7	6	2	86

gives a collective cost of 11, as shown in the last column of Table 6.3. The remaining actions can be classified by integrating information from Left-arm, Right-ankle, Left-thigh, and Waist nodes, as shown in Figure 6.8. Each one of these nodes can detect five, seven, four, and two actions, respectively.

Figure 6.8 shows the nodes required to identify each action using the constructed decision path. Visited nodes are listed along the x-axis, and actions are listed along the y-axis. The actions 4, 5, and 6—which are Sit to lie, Lie to sit, and Sit to lie

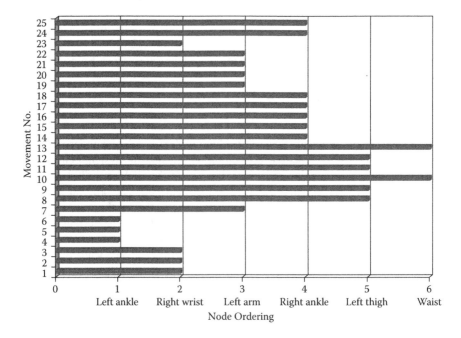

FIGURE 6.8   Identification order of actions using decision path. The path has a total cost of 86 and an average length of 3.44 per classification.

to sit—can be detected using only the Left-ankle node. This is in fact due to the unique patterns of the legs during transition between Sit and Lie postures. The actions that can be distinguished by the second node (Right-wrist) include actions 1, 2, 3, and 23. This implies that the actions that involve transitions between Stand and Sit can be identified by visiting one lower-body node (e.g., Left-ankle) and one upper-body node (Right-wrist). We note that the actions that involve Reach-up and Grasp (actions 7, 19, 20, 21, and 22) could not be classified using observations made by only one hand (Right-wrist). However, by visiting the third node (Left-arm), these actions can be distinguished from the rest. This can be interpreted by the fact that the set of Reach-up and Grasp actions listed in Table 6.2 includes both one-hand actions and two-hand actions, which require information from both hands to be distinguished from each other. Other actions that can be classified by visiting the three remaining nodes (Right-ankle, Left-thigh, and Waist) can be interpreted accordingly. From Figure 6.8, the average number of nodes required to detect an action is 3.44.

The greedy approximation described in Algorithm 2 is used to construct a full decision tree for the 25-action experiment. The resulting tree is shown in Figure 6.9. The order for visiting nodes in the full decision tree may change, depending on the action, because node ordering changes based on the target action. In the full decision tree illustrated in Figure 6.9, internal nodes, which correspond to the sensor nodes, are specified by light ellipsoids, while the leaves of the tree that represent actions are specified by dark squares. Each link in the tree is labeled by a number corresponding to the primitive that defines branch condition to the next node. A label ρ_{ir} on edge e_{ij} indicates the rth primitive within sensor node s_i that makes a branch to node s_j. The root node is the node whose local discrimination set (LDS) had maximum cardinality. The sensor node s_7 is considered to be the root node because LDS_7 had maximum cardinality (241) among all the nodes. When taking a branch, the new node is the sensor node capable of distinguishing the maximum number of remaining action pairs (a_k, a_l).

For each action, the path length is defined to be the number of internal nodes in the path from the root to that action. Figure 6.10 shows path length for each action using the decision path and the full decision tree. This value represents the number of sensor nodes required to detect each action. While the expected path length for the decision path is 3.44, the decision tree has an average value of 2.48, which indicates faster recognition compared with the linear ordering. Therefore, compared with the original system with nine sensor nodes, the decision-path and decision-tree models achieve 61.8% and 72.4% improvements in terms of node reduction. The cost of each decision tree is measured by summation of the path lengths over all actions. While the decision path has a cost of 86, the full tree has a total cost of 62.

6.6.3 Recognition Accuracy

The linear ordering of the nodes as well as the full decision tree are used to classify the test set (50% of the collected data). After filtering, segmentation, and feature extraction, each test trial is mapped to its corresponding primitives on each active node (one of ρ_{ir} clusters). The distributed algorithm described in Section 6.5.6 is

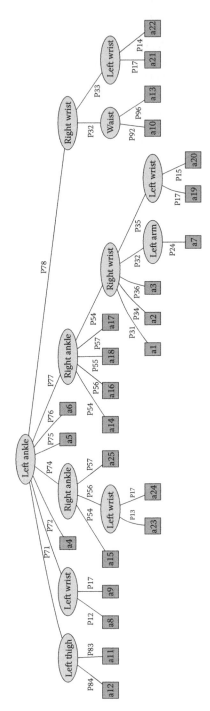

FIGURE 6.9   Full decision tree constructed to classify 25 actions. The tree has a total cost of 62 and an overall expected length of 2.48.

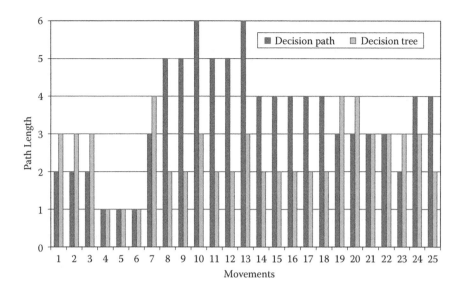

FIGURE 6.10 Path length for each action using linear ordering vs. full decision tree.

used to classify each unknown action as one of the predefined actions listed in Table 6.2. Using the primitive representation and the decision path defined by the ordering of active nodes, an accuracy of 95.7% is obtained. Using a full decision tree, an accuracy of 93.3% is achieved. Table 6.4 compares the effectiveness of the two decision-tree models against the original system with nine nodes. For the case of nine sensor nodes, it is assumed that features from all the nodes are used to build a k-NN (k-nearest neighbor) classifier at a base station. With the distributed classifier, the accuracy reduces by 2.5% and 4.9% using linear ordering and a branching decision tree, respectively.

6.6.4 Communication Saving

As mentioned previously, the decision-tree classifier reduces the communication cost by lowering the amount of data that is needed to be transmitted across the network. Figure 6.11 shows the amount of bandwidth required for classification of each action using centralized and distributed algorithms. The number associated with each action represents an instantaneous bandwidth that is calculated as the summation of bandwidths from all active nodes during occurrence of that action. In each classification scenario, the required bandwidth is a function of sampling frequency, the number of sensors, the number of nodes, and the size of data units that are transmitted. With a centralized classifier, each sensor node can transmit either raw sensor readings or a vector of statistical features to the base station. For transmitting raw data, it is assumed that each sensor reading is stored as a 12-bit value, which is sufficient for the readings acquired from the motion sensors. The 12-bit data unit is also enough to represent each feature.

TABLE 6.4

Comparing System Performance Before and After Each Optimization

Optimization	# of Active Nodes	Node Reduction (%)	Classification Algorithm	Classification Accuracy (%)
No optimization	9	0.0	k-NN (at base station)	98.2
Linear ordering	3.44	61.8	In-network (using decision path)	95.7
Branching tree	2.46	72.4	In-network (using full decision tree)	93.3

To estimate the bandwidth required for detection of each action, the IEEE 802.15.4 frame format is used, with the ratio of control to payload data being 0.2202. The control data include headers from both the physical (PHY) and media access control (MAC) layers. The following example explains how bandwidth is calculated for a particular action. For Stand to Sit, the average length of the action over all test trials is 123.9 samples, which translates into 2.48 s. For the case of transmitting raw data, each node needs to transmit $5 \times 123.9 \times 12 = 7,434$ bits over the 2.48-s period. Note that there is a total of five sensors in the discussed analysis, which are associated with x,y,z accelerometer and x,y gyroscope. The 7,434 bits of payload data can be sent through nine packets, each accommodating a maximum of 109 bytes of data. Thus, the total amount of data sent to the base station by each node is 7,434 +

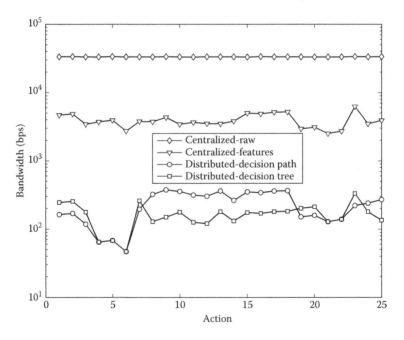

FIGURE 6.11 Comparison of communication savings of distributed classification versus centralized approaches.

(9 × 192) = 9,162 bits (assuming 24 bytes of control data per packet). Since all sensor nodes are involved in communication during a centralized approach, the total amount of data sent to the base station over the period of Stand to Sit (2.48 s) is 9,162 × 9 = 82,458 bits. This translates into 32.5-kbps bandwidth.

A similar approach can be used to estimate communication cost for other classification scenarios. A decision-tree-based classifier uses only a subset of the nodes to detect actions. Upon receiving the Target Movement Vector (TMV), each node updates the vector according to its local results and transmits the vector to the appropriate node in the tree. We note that the TMV is a binary vector that is sized according to the number of actions. Therefore, distributed classifiers require transmission of a 25-bit vector between the nodes. Obviously, the number of such transmissions varies from one action to another and depends on the number of active nodes involved in the classification. For example, only Left-ankle and Right-wrist nodes are involved in detection of Stand to Sit using a decision-path classifier (see Figure 6.8). The total amount of payload in this case is 25 bits, which needs to be transmitted in 2.48 s. A single packet can accommodate the 25 bits of payload. Adding the control data yields a total of 25 + 192 = 217 bits of data. Given that two nodes are involved in the classification and that each node broadcasts the TMV, this results in a bandwidth of 175 bps.

On average, the obtained bandwidths are 33.21 kbps, 3.91 kbps, 234 bps, and 167 bps for a centralized algorithm with raw data transmission, centralized algorithm with features transmission, decision path, and full decision tree, respectively.

6.7   CONCLUDING REMARKS

An important aspect of sensor-based motion analysis is robustness with respect to sensor displacement and misorientation. The accelerometer measures acceleration relative to the force of gravity. That is, the accelerometer readings that account for the actual dynamics of the actions are affected by gravity. However, in static mode, the accelerometer detects the gravity. This information is used to correct misplacement and misorientation of the nodes to maintain consistent mote positioning for all the experiments. While this approach helps in correcting initial node placements, more complex calibration models are needed to compensate for dynamic changes due to node misplacements during each action trial. One approach is to find signals/ features that are insensitive to node misplacement, and use them for classification [52]. Another technique builds a statistical model of misorientations, which approximates errors based on experimental data [53].

The phonological approach discussed in this chapter produces a compact expression of atomic actions. A clustering technique is usually used to generate basic primitives for each sensor node. The global state of the system is determined using the local knowledge given by primitives at each sensor node and a fast recognition policy. As observed in this chapter, the problem of determining the optimal ordering of sensor nodes for movement monitoring at the primitive level leads to constructing a minimum-cost decision tree. A full decision tree may further reduce the average number of sensor nodes required to detect each movement. The functionality of the discussed framework is verified using real data. Interestingly, the distributed classification technique achieves 95.7% and 93.3% accuracies in classifying human actions using decision path and full decision trees, respectively, with a 61.8% and 72.4% average reduction in node usage.

REFERENCES

1. Jovanov, E., A. Milenkovic, C. Otto, and P. de Groen. 2005. A wireless body area network of intelligent motion sensors for computer assisted physical rehabilitation. *J. NeuroEngineering Rehabilitation* 2 (1): 6.
2. Jones, W. 2007. Helmets sense the hard knocks [news]. *Spectrum, IEEE* 44 (10): 10–12.
3. Chen, J., K. Kwong, D. Chang, J. Luk, and R. Bajcsy. 2005. Wearable sensors for reliable fall detection. In *Engineering in Medicine and Biology Society (IEEE-EMBS 2005), 27th Annual International Conference of*, 3551–54. Piscataway, NJ: IEEE.
4. Morris, S., and J. Paradiso. 2002. Shoe-integrated sensor system for wireless gait analysis and real-time feedback. In *Engineering in Medicine and Biology: 24th Annual Conference and the Annual Fall Meeting of the Biomedical Engineering Society (EMBS/ BMES) Proceedings* 3:2468–69.
5. Pettersson, A., M. Engardt, and L. Wahlund. 2002. Activity level and balance in subjects with mild Alzheimer's disease. *Dementia Geriatric Cognitive Disorders* 13:213–16.
6. Mitchell, S. L., J. J. Collin, C. J. D. Luca, A. Burrows, and L. A. Lipsitz. 1995. Open-loop and closed-loop postural control mechanisms in Parkinson's disease: Increased mediolateral activity during quiet standing. *Neuroscience Letters* 197 (2): 133–36.
7. Emmerik, V., and V. Wegen. 2002. On the functional aspects of variability in postural control. *Exercise and Sport Sciences Reviews* 30 (4): 177–83.
8. Gelb, D., E. Oliver, and S. Gilman. 1999. Diagnostic criteria for Parkinson disease. *Arch. Neurol.* 56:33–39.
9. Raghunathan, V., C. Schurgers, S. Park, and M. Srivastava. 2002. Energy-aware wireless microsensor networks. *IEEE Signal Processing Magazine* 19 (2): 40–50.
10. Hu, W., T. Tan, L. Wang, and S. Maybank. 2004. A survey on visual surveillance of object motion and behaviors. *IEEE Trans. Systems, Man, and Cybernetics, Part C: Applications and Reviews* 34 (3): 334–52.
11. Patterson, D. J., D. Fox, H. Kautz, and M. Philipose. 2005. Fine-grained activity recognition by aggregating abstract object usage. In *ISWC '05: Proceedings of the Ninth IEEE International Symposium on Wearable Computers*, 44–51. Washington, DC: IEEE Computer Society.
12. Philipose, M., K. P. Fishkin, M. Perkowitz, D. J. Patterson, D. Fox, H. Kautz, and D. Hahnel. 2004. Inferring activities from interactions with objects. *IEEE Pervasive Computing* 3 (4): 50–57.
13. Vehkaoja, A., M. Zakrzewski, J. Lekkala, S. Iyengar, R. Bajcsy, S. Glaser, S. Sastry, and R. Jafari. 2007. A resource optimized physical movement monitoring scheme for environmental and on-body sensor networks. In *HealthNet '07: Proceedings of the 1st ACM SIGMOBILE International Workshop on Systems and Networking Support for Healthcare and Assisted Living Environments*, 64–66. New York: ACM Press.
14. Jafari, R., F. Dabiri, P. Brisk, and M. Sarrafzadeh. 2005. Adaptive and fault tolerant medical vest for life-critical medical monitoring. In *SAC '05: Proceedings of the 2005 ACM Symposium on Applied Computing*, 272–79. New York: ACM Press.
15. Logan, B., J. Healey, M. Philipose, E. Tapia, and S. Intille. 2007. A long-term evaluation of sensing modalities for activity recognition. *Lecture Notes in Computer Science* 4717:483.
16. Wu, W. H., A. A. T. Bui, M. A. Batalin, L. K. Au, J. D. Binney, and W. J. Kaiser. 2008. MEDIC: Medical embedded device for individualized care. *Artif. Intell. Med.* 42 (2): 137–52.

17. Duda, R. O., P. E. Hart, and D. G. Stork. 2000. *Pattern classification.* New York: Wiley-Interscience.
18. Klingbeill, L., and T. Wark. 2008. A wireless sensor network for real-time indoor localisation and motion monitoring. In *IPSN '08: Proceedings of the 7th International Conference on Information Processing in Sensor Networks*, 39–50. Washington, DC: IEEE Computer Society.
19. Bao, L., and S. S. Intille. 2004. Activity recognition from user-annotated acceleration data. In *Pervasive 2004 Conference Proceedings*, 1–17. Berlin: Springer-Verlag. http://dx.doi.org/10.1007/b96922.
20. Lester, J., T. Choudhury, and G. Borriello. 2006. A practical approach to recognizing physical activities. In *Pervasive 2006 Conference Proceedings*, 1–16. Berlin: Springer-Verlag. http://dx.doi.org/10.1007/11748625_1.
21. Duda, R., P. Hart, and D. Stork. 1973. *Pattern classification and scene analysis.* New York: Wiley.
22. Ghasemzadeh, H., E. Guenterberg, and R. Jafari. 2009. Energy-efficient information-driven coverage for physical movement monitoring in body sensor networks. *IEEE J. Selected Areas in Commun.* 27:58–69.
23. Jolliffe, I. 2002. *Principal component analysis.* 2nd ed. New York: Springer.
24. Guerra-Filho, G., C. Fermüller, and Y. Aloimonos. 2005. Discovering a language for human activity. In *Proceedings of the AAAI 2005 Fall Symposium on Anticipatory Cognitive Embodied Systems*, 70–77. Washington, DC: AAAI.
25. Polastre, J., R. Szewczyk, and D. Culler. 2005. Telos: Enabling ultra-low power wireless research. In *Proceedings of Fourth International Symposium on Information Processing in Sensor Networks (IPSN '05)*, 364–69. Piscataway, NJ: IEEE Press.
26. Stergiou, N., ed. 2003. *Innovative analyses of human movement: Analytical tools for human movement research.* Champaign, IL: Human Kinetics.
27. Ghasemzadeh, H., and R. Jafari. 2011. Physical movement monitoring using body sensor networks: A phonological approach to construct spatial decision trees. *IEEE Trans. Industrial Informatics* 7 (1): 66–77.
28. Guenterberg, E., S. Ostadabbas, H. Ghasemzadeh, and R. Jafari. 2009. An automatic segmentation technique in body sensor networks based on signal energy. In *BodyNets '09, Proceedings of the Fourth International Conference of Body Area Networks*, 21. New York: ACM Press.
29. Ward, J., P. Lukowicz, G. Tröster, and T. Starner. 2006. Activity recognition of assembly tasks using body-worn microphones and accelerometers. *IEEE Trans. Pattern Analysis and Machine Intelligence* 28 (10): 1553–67.
30. Guenterberg, E., H. Ghasemzadeh, and R. Jafari. 2009. A distributed hidden Markov model for fine-grained annotation in body sensor networks. In *Wearable and Implantable Body Sensor Networks (BSN '09), Sixth International Workshop on*, 339–44. Piscataway, NJ: IEEE.
31. Guenterberg, E., H. Ghasemzadeh, V. Loseu, and R. Jafari. 2009. Distributed continuous action recognition using a hidden Markov model on body sensor networks. *Distributed Computing in Sensor Systems* 5516:145–58.
32. Jafari, R., R. Bajcsy, S. Glaser, B. Gnade, M. Sgroi, and S. Sastry. 2007. Platform design for health-care monitoring applications. In *High Confidence Medical Devices, Software, and Systems and Medical Device Plug-and-Play Interoperability (HCMDSS-MDPnP), Joint Workshop on*, 88–94. Piscataway, NJ: IEEE.
33. Lymberopoulos, D., A. Ogale, A. Savvides, and Y. Aloimonos. 2006. A sensory grammar for inferring behaviors in sensor networks. In *Proceedings of the Fifth International Conference on Information Processing in Sensor Networks*, 251–59. New York: ACM Press.

34. Rabiner, L. 1989. A tutorial on hidden Markov models and selected applications in speech recognition. *Proceedings of the IEEE* 77 (2): 257–86.

35. Aggarwal, J., and S. Park. 2004. Human motion: Modeling and recognition of actions and interactions. In *3D Data Processing, Visualization and Transmission (3DPVT 2004), Proceedings of 2nd International Symposium on*, 640–47. Piscataway, NJ: IEEE.

36. Niwase, N., J. Yamagishi, and T. Kobayashi. 2005. Human walking motion synthesis with desired pace and stride length based on HSMM. *IEICE Trans. Inf. Syst.* E88-D (11): 2492–99.

37. Guerra-Filho, G., C. Fermuller, and Y. Aloimonos. 2005. Discovering a language for human activity. In *Proc. of the AAAI 2005 Fall Symposium on Anticipatory Cognitive Embodied Systems (FS '05)*, 70–77. Washington, DC: AAAI.

38. Reng, L., T. B. Moeslund, and E. Granum. 2006. Finding motion primitives in human body gestures. In *GW Proceedings of the 6th International Conference on Gesture in Human-Computer Interaction and Simulation*. Berlin: Springer-Verlag, 133–44.

39. Johnson, S. C. 1967. Hierarchical clustering schemes. *Psychometrika* 32 (3): 241–54.

40. MacQueen, J. B. 1967. Some methods for classification and analysis of multivariate observations. In *Proc. of the Fifth Berkeley Symposium on Mathematical Statistics and Probability*, vol. 1, ed. L. M. L. Cam and J. Neyman, 281–97. Berkeley: University of California Press.

41. Berkhin, P. 2006. A survey of clustering data mining techniques. In *Grouping Multidimensional Data*. Berlin: Springer, 25–71.

42. Steinley, D., and M. J. Brusco. 2007. Initializing K-means batch clustering: A critical evaluation of several techniques. *J. Classif.* 24 (1): 99–121.

43. Lamrous, S., and M. Taileb. 2006. Divisive hierarchical K-means. In *Computational Intelligence for Modelling, Control and Automation, and International Conference on Intelligent Agents, Web Technologies and Internet Commerce*, 18–36. Piscataway, NJ: IEEE.

44. Rousseeuw, P. 1987. Silhouettes: A graphical aid to the interpretation and validation of cluster analysis. *J. Comput. Appl. Math.* 20 (1): 53–65.

45. Breu, H., J. Gil, D. Kirkpatrick, and M. Werman. 1995. Linear time Euclidean distance transform algorithms. *Pattern Analysis and Machine Intelligence, IEEE Trans.* 17 (5): 529–33.

46. Ghasemzadeh, H., J. Barnes, E. Guenterberg, and R. Jafari. 2008. A phonological expression for physical movement monitoring in body sensor networks. In *Mobile Ad Hoc and Sensor Systems (MASS 2008), 5th IEEE International Conference on*, 58–68. Piscataway, NJ: IEEE.

47. Sieling, D. 2008. Minimization of decision trees is hard to approximate. *J. Comput. Syst. Sci.* 74 (3): 394–403.

48. Feige, U., and P. Tetali. 2004. Approximating min sum set cover. *Algorithmica* 40 (4): 219–34.

49. Heeringa, B. 2006. Improving access to organized information. PhD diss., UMass Amherst.

50. Chakaravarthy, V. T., V. Pandit, S. Roy, P. Awasthi, and M. Mohania. 2007. Decision trees for entity identification: Approximation algorithms and hardness results. In *PODS '07: Proceedings of the twenty-sixth ACM SIGMOD-SIGACT-SIGART symposium on principles of database systems*, 53–62. New York: ACM Press.

51. Graham, R. L., B. L. Rothschild, and J. H. Spencer. 1980. *Ramsey theory*. New York: Wiley-Interscience.

52. Harms, H., O. Amft, and G. Tröster. 2009. Modeling and simulation of sensor orientation errors in garments. In *Bodynets 2009: Proceedings of the 4th International Conference on Body Area Networks*. New York: ACM Press.

53. Kunze, K., and P. Lukowicz. 2008. Dealing with sensor displacement in motion-based onbody activity recognition systems. In *UbiComp '08: Proceedings of the 10th International Conference on Ubiquitous Computing*, 20–29. New York: ACM Press.

7 A Network Structure for Delay-Aware Applications in Wireless Sensor Networks

Chi-Tsun Cheng, Chi K. Tse, and Francis C. M. Lau

CONTENTS

7.1 INTRODUCTION

Wireless sensor networks comprise large numbers of wireless sensor nodes. The compact design of wireless sensor nodes allows them to fit in most environments and perform close-range sensing. Such a unique feature makes wireless sensor networks highly suitable for monitoring in extreme conditions. In most sensing applications, sensing information should be sampled regularly and spatially over a period of time. Wireless sensor nodes are battery-powered devices. To maintain the required sensing coverage over such a period, wireless sensor nodes should conserve their energy aggressively while remaining operational. Much prior work has focused on conserving energy by clustering. A network with clustering is divided into several clusters. Within each cluster, one of the sensor nodes is elected as a *cluster head* (CH), with the rest being *cluster members* (CMs). The CH will collect data from its CMs directly or in a multihop manner. By organizing wireless sensor nodes into clusters, energy dissipation is reduced by decreasing the number of nodes involved in long-distance transmission [1]. The number of data transmissions and the amount of energy consumption can be further reduced by performing data/decision fusion on nodes along

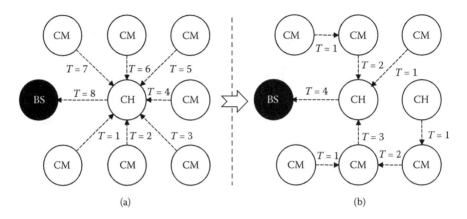

FIGURE 7.1 (a) Data collection in a two-hop network and (b) data collection in an improved multihop network. Circles with CM represent the cluster members. Circles with CH represent the cluster heads. Filled circles with BS represent the base stations. A dashed arrow represents the existence of a data link, and the direction of the arrow shows the direction of data flow.

the data-aggregation path. Clustering provides a significant improvement in energy savings.

In sensor networks with clusters, it is common for a CH to collect data from its CMs one by one. Let T be the average transmission delay among nodes. Data packets generated by sensor nodes are considered as highly correlated, and thus a node is always capable of fusing all received packets into a single packet by means of data/decision-fusion techniques [2, 3]. Referring to the situation shown in Figure 7.1a, a *base station* (BS) will take $8 \times T$ to collect a complete set of data from the network. By transforming the network into a multihop network, as shown in Figure 7.1b, it can be shown that the time needed by the BS to collect a full set of data from the network can be reduced to $4 \times T$. In the modified network, apart from requiring a shorter delay in data collection, CMs will need smaller buffers to handle the incoming data while waiting for the associated CH to become available.

The aim of this chapter is to investigate the characteristics of a delay-aware data collection network structure in wireless sensor networks. An algorithm for forming such a network structure is proposed for different scenarios. The proposed algorithm is operating between the data link layer and the network layer. The algorithm will form networks with minimum delays in the data collection process. At the same time, the algorithm will try to keep the transmission distance among wireless sensor nodes at low values in order to limit the amount of energy consumed in communications.

The rest of the chapter is organized as follows: Section 7.2 briefly reviews related work. Section 7.3 defines the proposed network structure. Section 7.4 explains the algorithm for forming the proposed network structure in different scenarios. A numerical analysis is given in Section 7.5 to show how different network structures perform in terms of delays in data collection processes. Simulation results and their analysis are given in Sections 7.6 and 7.7, respectively. Finally, the chapter is concluded in Section 7.8.

7.2   RELATED WORK

Due to the energy constraints of individual sensor nodes, energy conservation becomes one of the major issues in sensor networks. In wireless sensor networks, a large portion of the energy in a node is consumed in wireless communications. The amount of energy consumed in a transmission is proportional to the corresponding communication distance. Therefore, long-distance communications between nodes and the BS are usually not encouraged. One way to reduce energy consumption in sensor networks is to adopt a clustering algorithm [1]. A clustering algorithm tries to organize sensor nodes into clusters. Within each cluster, one node is elected as the CH. The CH is responsible for (1) collecting data from its CMs, (2) fusing the data by means of data/decision-fusion techniques, and (3) reporting the fused data to the remote BS. In each cluster, the CH is the only node involved in long-distance communications. Energy consumption of the whole network is therefore reduced.

Intensive research [2–5] has been conducted on reducing energy consumption by forming clusters with appropriate network structures. Heinzelman, Chandrakasan, and Balakrishnan [2] proposed a clustering algorithm called LEACH. In networks using LEACH, sensor nodes are organized in multiple-cluster two-hop (MC2H) networks (i.e., CMs→CH→BS). Using the idea of clustering, the amount of long-distance transmissions can be greatly reduced. Lindsey and Raghavendra [3] proposed another clustering algorithm called PEGASIS, which is a completely different approach that organizes sensor nodes into a single-chain (SC) network. In such networks, a single node on the chain is selected as the CH. By minimizing the number of CHs, the energy consumed in long-distance transmission is further minimized. Tan and Körpeoğlu [4] developed PEDAP, which is based on the idea of a minimum spanning tree (MST). Besides minimizing the amount of long-distance transmission, the communication distances among sensor nodes are minimized. Fonseca et al. [5] proposed the collection tree protocol (CTP). The CTP is a kind of gradient-based routing protocol that uses *expected transmissions* (ETX) as its routing gradient. ETX is the number of expected transmissions of a packet necessary for it to be received without error [6]. Paths with low ETX are expected to have high throughput. Nodes in a network using CTP will always pick a route with the lowest ETX. In general, the ETX of a path is proportional to the corresponding path length [7]. Thus, CTP can greatly reduce the communication distances among sensor nodes. All these algorithms show promising results in energy saving. However, a network formed by an energy-efficient clustering algorithm may not necessarily be desirable for data collection. An analysis on how these network structures perform in terms of data collection efficiency is given in Section 7.5.

The focus of this chapter is on investigating the data collection efficiency of networks formed by different clustering algorithms. Therefore, event-triggering algorithms such as TEEN [8] and APTEEN [9] will not be considered in this chapter. A related work on data collection efficiency was done by Florens, Franceschetti, and McEliece [10]. In their work, lower bounds on data collection time are derived for various network structures. However, the effect of data fusion, which is believed to be one of the major features of sensor networks, was not considered. Wang et al. [11] proposed link-scheduling algorithms for wireless sensor networks, which can raise network throughput considerably. In their work, however, it is assumed that data links among wireless sensor nodes are predefined. In contrast, the objective of this chapter is to form data links among wireless sensor nodes and thus to shorten the delays in the data collection processes. Another related work was done by Solis and Obraczka [12], who studied the impact of timing in data aggregation for sensor networks. Chen and Wang [13] investigated the effects of network capacity under different network structures and routing strategies. A similar work was done by Song and He [14]. In their work, the term *capacity* is defined as the maximum end-to-end traffic that a network can handle. The delay in a data collection process is not their major concern.

7.3 THE PROPOSED NETWORK STRUCTURE

The proposed network structure is a tree structure. To deliver the maximum data collection efficiency, the number of nodes N in the proposed network structure has to be restricted to $N = 2^p$, where $p = 1, 2, \ldots$. It will be shown in a later part that such restriction can be relaxed by giving up some performance. Each CM will be given a rank, which is an integer between 1 and p. A node with rank k will form $k - 1$ data links with $k - 1$ nodes, while these $k - 1$ nodes are with different ranks starting from 1, 2,..., up to $k - 1$. All these $k - 1$ nodes will become the child nodes of the node with rank k. The node with rank k will form a data link with a node with a higher rank. This higher rank node will become the parent node of the node with rank k. The CH will be considered as a special case. The CH is the one with the highest rank in the network. Instead of forming a data link with a node of higher rank, the CH will form a data link with the BS. By following this logic, the distribution of the rank will follow an inverse exponential base-2 function, as shown in Table 7.1. An example of the proposed network with $N = 16$ is shown in Figure 7.2. In this example, it takes $5 \times T$ for the BS to collect all data from 16 nodes. By dividing the time domain into time slots of durations T, the process discussed in this example will last for five time slots.

Lemma 1. Consider a network with $N = 2^p$, where $p = 1, 2, \ldots$. Data packets generated by sensor nodes are considered as highly correlated, and thus a node is always capable of fusing all received packets into a single packet by means of data/decision fusion techniques. Through adopting the proposed network structure, a node i of rank $k \geq 2$ (where $k \subset Z$) requires $k - 1$ time slots to collect data from all its child nodes.

Proof. Consider a network with $N = 2^p$, where $p = 1, 2, \ldots$. For a node of rank $k = 2$, the time slots required for it to collect data from all its child nodes is equal to the number

---

TABLE 7.1

CMs' Rank Distribution in the Proposed Network Structure with Network Size $N = 2^p$, where $p = 1, 2, \ldots$

Rank	1	2	...	$\log_2 N - 1$	$\log_2 N$
No. of nodes	$\dfrac{N}{2^1}$	$\dfrac{N}{2^2}$	...	$\dfrac{N}{2^{(\log_2 N - 1)}}$	$\dfrac{N}{2^{(\log_2 N)}}$

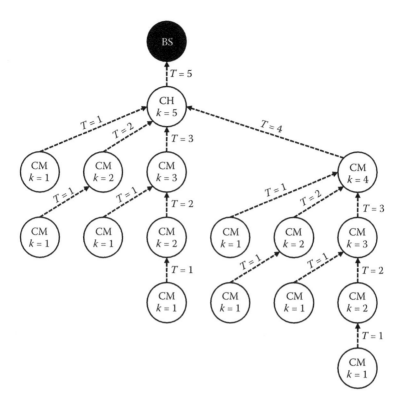

FIGURE 7.2 Proposed network structure with network size $N = 16$. Circles with CM represent the cluster members. The circle with CH represents the cluster head. The filled circle with BS represents the base station. The rank of each node is represented by the variable k. A dashed arrow represents the existence of a data link, and the direction of the arrow shows the direction of data flow.

of child nodes it has, which is 1. Thus the case for $k = 2$ is true. Now let us assume that any node of connection $k = n$ requires $n - 1$ time slots to collect all data from its child nodes. For node i of rank $k = n + 1$, it has n directly connected child nodes. Each of these directly connected child nodes has different ranks ranging from 1 to n. Thus, they need 0 to $n - 1$ time slots to collect data from all their subchild nodes plus one extra time slot to report their aggregated data to node i. Therefore, the maximum time slots required for node i to collect data from all its child nodes is $k - 1 = n$. By induction, Lemma 1 is proved. ∎

Theorem 1. Consider a network with $N = 2^p$, where $p = 1, 2, \ldots$. Data packets generated by sensor nodes are considered as highly correlated, and thus a node is always capable of fusing all received packets into a single packet by means of data/decision-fusion techniques. By adopting the proposed network structure, the number of time slots $t(N)$ required for the BS to collect data from the whole network is given by

$$t(N) = \log_2 N + 1. \tag{7.1}$$

Proof. Consider a network with $N = 2^p$, where $p = 1,2,\ldots$. Through adopting the proposed network structure, the CH is the only node with the highest ranking, which is

$$k_{max} = \log_2 N + 1.$$

From Lemma 1, the number of time slots $t(N)$ required for a CH, with rank k_{max}, to collect data from all its child nodes is

$$t(N) = k_{max} - 1 = \log_2 N.$$

Thus, the number of time slots $t(N)$ required for the BS to collect data from the whole network is the time slots required by the CH to collect data from all its child nodes plus one, i.e.,

$$t(N) = \log_2 N + 1. \qquad \blacksquare$$

7.4 NETWORK-FORMATION ALGORITHM

It has been proven in the last section that the delay in the data collection process of a wireless sensor network can be greatly reduced by adopting the proposed network structure. Since energy consumption is always a major issue in the study of wireless sensor networks, the objective of the proposed network-formation algorithms is, therefore, to achieve the proposed network structure while keeping the energy consumption in the data collection process at a low value.

A wireless sensor node can be considered as a device built up of three major units, namely the microcontroller unit (MCU), the transceiver unit (TCR), and the sensor board (SB). Each of these units will consume a certain amount of energy while operating. The energy consumed by a wireless sensor node i can be expressed as

$$E_{i\_SN} = E_{i\_MCU} + E_{i\_TCR} + E_{i\_SB}, \tag{7.2}$$

where E_{i_MCU} represents the energy consumed by the MCU, E_{i_TCR} represents the energy consumed by the TCR, and E_{i_SB} represents the energy consumed by the SB. Here, E_{i_TCR} can be further expressed as

$$E_{i\_TCR} = E_{i\_TCR\_RX} + E_{i\_TCR\_TX}(d_i), \tag{7.3}$$

where $E_{i\_TCR\_RX}$ denotes the energy consumed by the TCR in receiving mode, while $E_{i\_TCR\_TX}(d_i)$ denotes the energy consumed by the TCR to transmit for a distance of d_i. The total energy consumed by a network of N sensor nodes is expressed as

$$E_{TOT}(N) = \sum_{i=1}^{N} \Big(E_{i\_MCU} + E_{i\_TCR\_RX}$$
$$+ E_{i\_TCR\_TX}(d_i) + E_{i\_SB} \Big). \tag{7.4}$$

Normally, E_{i_MCU}, E_{i_TCR}, and $E_{i\_TCR\_RX}$ are constants. On the other hand, $E_{i\_TCR\_TX}(d_i)$ is a function of d_i, which is heavily dependent on the network structure. Therefore, Equation (7.4) can be simplified as follows:

$$E_{TOT}(N) = C_1 + \sum_{i=1}^{N} E_{i\_TCR\_TX}(d_i),\qquad(7.5)$$

where C_1 is a constant. Assume that the path loss exponent is equal to 2, $E_{i\_TCR\_TX}(d_i)$ can be further expressed as

$$E_{i\_TCR\_TX}(d_i) = E_{i\_TCR\_EC} + E_{i\_TCR\_PA}d_i^2,\qquad(7.6)$$

where $E_{i\_TCR\_EC}$ is the energy consumed by the TCR's electronic circuitry, while $E_{i\_TCR\_PA}$ denotes the energy consumed by the power amplifier of the TCR. Both $E_{i\_TCR\_EC}$ and $E_{i\_TCR\_PA}$ are constants and, therefore, Equation (7.5) can be expressed as

$$E_{TOT}(N) = C_1 + C_2 + C_3 \sum_{i=1}^{N} d_i^2,\qquad(7.7)$$

where C_2 and C_3 are constants. Here, Equation (7.7) shows that the total energy consumption of the network can be minimized by reducing $\sum_{i=1}^{N} d_i^2$. Thus, the objective of the proposed network-formation algorithms is to construct the proposed network structure while keeping $\sum_{i=1}^{N} d_i^2$ at a low value.

In this section, a network-formation algorithm is proposed to achieve this objective. Basically, the operation of the proposed network-formation algorithm is to join clusters of the same size together. It can be implemented in either a centralized or decentralized fashion. Specifically, the decentralized version can be described as follows:

1. Each node is labeled with a unique identity and marked as level w. The unique identity will serve only as an identification, which has no relation with the sensor nodes' locations and connections. Here, w is a function that represents the number of nodes in a cluster. For a cluster of i nodes, its w value is equal to $\log_2 i$. Since nodes are disconnected initially (i.e., no data link exists among wireless sensor nodes), these N nodes can be considered as N level-0 clusters. Within each cluster, one node will be elected as the sub-CH. We denote SCH(w) as a sub-CH of a level-w cluster. In the proposed algorithm, an SCH can only make connection (i.e., set up a data link) with another SCH of the same level. Because there is only one node in each cluster, all nodes begin as SCH(0). The dimensions of the terrain (t_x, t_y) are provided to the sensor nodes before deployment.
2. Each SCH performs random back-off and then broadcasts a *density probing packet* (DPP) to its neighboring SCHs, which are within a distance of $r_{dp} = (t_x^2 + t_y^2)^{\frac{1}{2}}$ m. Note that the size of a DPP is much smaller than that of a

data packet. An SCH can use the number of received DPPs, together with the dimensions of the terrain, to estimate the total number of nodes (N_{est}) in the network. An SCH will use the N_{est} to adjust its communication distance r_{com}. Definition of r_{com} will be explained later in this section.

3. Each SCH will do a random back-off and then broadcast an *invitation packet* (IVP) to its neighbors within r_{com} m. The IVP contains the level w and the identity of the issuing SCH. An SCH will estimate the distances to its neighboring SCHs using the received signal strength of the IVPs. An SCH will also count the number of IVPs received. If the number of IVPs received has exceeded a predefined threshold or if a maximum duration has been reached, an SCH will send a *connection request* (CR) to this nearest neighbor. If both SCHs are the nearest neighbor of each other, a connection will be formed between these two SCHs.

4. Once they are connected, the two SCHs and their associated level-w clusters will form a composite level-$w + 1$ cluster. One of the two involved SCHs will become the chief SCH of the composite cluster. The chief SCH will listen to the communication channel and reply to any CRs from lower levels with a *rejecting packet* (RP). When no more CRs from lower levels can be heard, the chief SCH will start to make connection with other SCHs of the same level.

5. If an RP is received, an SCH will send a CR to its next nearest neighbor in its database. If such a neighbor does not exist, the SCH will increase its r_{com}. The SCH will then broadcast a CR using the new r_{com}. Upon receiving the CR, an SCH of the same level will grant the request if it is still waiting for a CR.

6. If no connection can be made within a period of time, because either all neighbors of the same level are unavailable or all CRs have been rejected, the SCH will increase its r_{com} and broadcast the CR again. This process repeats as long as $r_{com} < \sqrt{t_x^2 + t_y^2}$. If $r_{com} = \sqrt{t_x^2 + t_y^2}$, the SCH will make connection with the BS directly.

7. These processes continue until no more connection can be formed.

In the proposed algorithm, the communication distance r_{com} is defined as

$$r_{com} = \frac{\sqrt{t_x^2 + t_y^2}}{\alpha - \beta - w}, \quad \beta + w < \alpha. \tag{7.8}$$

Here, β is a constant that is set to 0 initially. Parameter α is the estimated maximum rank of a node in the network, which is expressed as

$$\alpha = \lceil \log_2(N_{est}) \rceil + 1 \tag{7.9}$$

Initially, all SCHs are with $w = 0$ and $\beta = 0$. Therefore, the SCHs will start broadcasting their IVPs with $r_{com} = (\sqrt{t_x^2 + t_y^2})(\alpha - 0 - 0)^{-1}$. If an SCH has made a connection with another SCH, its level will be increased by 1 (i.e., $w = 1$). After that, the chief SCH of the composite cluster will broadcast its IVP with $r_{com} = (\sqrt{t_x^2 + t_y^2})(\alpha - 0 - 1)^{-1}$. The r_{com} is designed to be increased with w because when SCHs are paired up to form composite clusters, the average separation among composite clusters will be increased. It is more energy efficient to start the broadcasting with a longer communication range. However, if no connection can be made, an SCH will increase its β by 1. This will increase r_{com}, which can facilitate the searching of available SCHs. An SCH will increase its r_{com} by incrementing β until a connection can be made. The sum of β and w is defined to be smaller than α to ensure that r_{com} is upper-bounded by the diagonal of the sensing terrain.

In step 3 of the proposed algorithm, an SCH will send a *connection request* (CR) to its nearest neighbor if the number of received IVPs has exceeded $\aleph$. Here, $\aleph$ is the expected number of IVPs to be received, which is expressed as

$$\aleph = \left\lceil \frac{\eta \pi r_{com}^2 - 1}{w} \right\rceil. \tag{7.10}$$

Parameter η is the density of the network, which can be estimated using the N_{est} obtained previously.

When being implemented in a decentralized control manner, the proposed algorithm may end up with multiple composite clusters if the number of nodes is not equal to 2^p, where $p = 1,2,\ldots$. SCHs of these composite clusters will communicate with the BS directly. By virtue of pairing up composite clusters of same sizes, the algorithm will end up with composite clusters of completely different sizes. Considering the BS as the root of the network, the number of time slots required by the BS to collect data from all sensor nodes is

$$t(N) = \lceil \log_2 (N+1) \rceil. \tag{7.11}$$

In contrast, the proposed algorithm can also be carried out at the BS as a centralized control algorithm. The BS is assumed to have the coordinates of all sensor nodes in the network. When the number of nodes is not equal to 2^p, where $p = 1,2,\ldots$, dummy nodes can be virtually added in the calculation process, depending on the application. If a single cluster is required, dummy nodes should be virtually added to fulfill the requirement of $N = 2^p$, where $p = 1,2,\ldots$. Here, dummy nodes are not physical nodes, but virtual nodes used to facilitate the computation at the BS. If multiple clusters can be formed, dummy nodes are not essential. When dummy nodes are virtually added, these dummy nodes will have infinite separations from the real nodes and from themselves. Note that whenever a real SCH is connecting with a dummy SCH, the real SCH will always be the chief SCH of the composite cluster. This is to ensure that the removal of dummy nodes at the end of the calculation process will not partition the network. The number of time slots required by the BS to collect data from all sensor nodes will be governed by Equation (7.1) (for single cluster) Equation (7.11) (for multiple clusters).

7.5 NUMERICAL ANALYSIS

Theorem 2. Assume that each sensor node can only communicate with one sensor node at a time, and that data fusion is applicable. For a single-cluster network of $N = 2^p$ nodes, where $p = 1,2,\dots$, the minimum number of time slots required by the BS to collect data from N sensor nodes is

$$t(N)_{min} = \log_2 N + 1. \tag{7.12}$$

Proof. Given a period of t time slots, a parent node v can collect data from, at most, t directly connected child nodes, provided that these t child nodes are using different time slots to communicate. Within these t child nodes, the uth node will report data at time slot u, which implies that the uth node can collect data from at most $u - 1$ directly connected child nodes of itself before it has to report data to its parent node. Therefore, for a period of t time slots, a parent node v can receive data from at most 2^t nodes (including itself). On the other hand, the minimum number of time slots required for a parent node to collect data from N nodes (including itself) is $\log_2 N$. Thus, the minimum number of time slots required for a BS to collect data from N nodes is $t(N)_{min} = \log_2 N + 1$. ∎

For a single-cluster network with N nodes, where $N > 0|N \subset Z$, the minimum number of time slots required for a BS to collect data from N nodes is

$$t(N)_{min} = \lceil \log_2 N \rceil + 1. \tag{7.13}$$

From Theorem 1 and Equation (7.1), it can be shown that the proposed network structure is an optimum network structure in terms of data collection efficiency provided that

1. Each sensor node can only communicate with one sensor node at a time.
2. Data fusion can be carried out at every sensor node.
3. Sensor nodes belong to a single cluster with a single CH.

The same idea can be applied to a multiple-cluster network by considering the BS as the root of the network structure. Therefore, in multiple-cluster networks, the minimum number of time slots required for a BS to collect data from N nodes is

$$t(N)_{min} = \lceil \log_2(N + 1) \rceil. \tag{7.14}$$

Using Equation (7.11), it can be shown that the proposed network structure is again an optimum network structure in terms of data collection efficiency provided that

1. Each sensor node can only communicate with one sensor node at a time.
2. Data fusion can be carried out at every sensor node.
3. The network consists of multiple clusters.

In an MC2H network with N nodes organized in g clusters, where $N \geq \Sigma_{m=1}^{g} m$, the time slots required by the BS to collect data from all sensor nodes is minimized when all clusters have different numbers of nodes, as this means that each cluster can communicate with the BS in an interleaved manner. Meanwhile, the number of nodes in the largest cluster should be minimized such that the total number of time slots required by the BS is also minimized. An example for $g = 2$ is shown as follows.

Example 7.1

For an MC2H network of N nodes organized in two clusters (where $N \geq 3$), in order to achieve the maximum data collection efficiency, the number of nodes in these two clusters should be equal to

$$\frac{N+1}{2} \text{ and } \frac{N-1}{2}, \text{ for } N \text{ odd}$$

$$\frac{N}{2}+1 \text{ and } \frac{N}{2}-1, \text{ for } N \text{ even.}$$

(7.15)

The minimum number of time slots $t(N)_{min}$ required by the BS to collect data from all sensor nodes is equal to the number of nodes in the largest cluster. Therefore,

$$t(N)_{min} = \begin{cases} \dfrac{N+1}{2}, & N \text{ is odd} \\ \dfrac{N}{2}+1, & N \text{ is even.} \end{cases}$$

(7.16)

In general, for an MC2H network of N nodes organized in g clusters, where $N \geq \Sigma_{m=1}^{g} m$, the number of nodes in the jth cluster can be written as

$$\left\lfloor \frac{N - S_g + (j-1)(g+1)}{g} \right\rfloor + 1, j = 1, 2, 3, \ldots g$$

(7.17)

where $\lfloor u \rfloor$ denotes the nearest integer smaller than u and $S_g = \Sigma_{m=1}^{g} m$. Thus, the minimum number of time slots $t(N)_{min}$ required by the BS to collect data from all sensor nodes is equal to

$$t(N)_{min} = \left\lfloor \frac{N - S_g + (g-1)(g+1)}{g} \right\rfloor + 1.$$

(7.18)

Based on Equation (7.17), the optimum number of clusters g_{opt} for an MC2H network in terms of data collection efficiency can be obtained from the following inequality:

$$\frac{(1+g)g}{2} \geq N$$

$$\Rightarrow \qquad g \geq \frac{-1+\sqrt{1^2+8N}}{2}$$

(7.19)

$$\Rightarrow \qquad g_{opt} = \left\lceil \frac{-1+\sqrt{1^2+8N}}{2} \right\rceil,$$

(7.20)

where N is the number of nodes in the network, and g is the number of clusters.

Theorem 3. For an MC2H network of N nodes organized in g clusters of completely different sizes, where $g \leq N < S_g$, the minimum number of time slots $t(N)_{\min}$ required by the BS to collect data from all sensor nodes is equal to the number of clusters in the system, i.e., g.

Proof. Consider the extreme case, in which an MC2H network of N nodes is organized in g clusters of completely different sizes, where $N = S_g$. These g clusters will all have different numbers of nodes ranging from 1 to g. The minimum number of time slots $t(N)_{\min}$ required by the BS to collect data from all sensor nodes is equal to the number of nodes in the largest cluster, which is g. Suppose that one node has to be removed from the network such that N is reduced to $N - 1$. To maintain the number of clusters in the network, this particular node must be removed from one of the clusters except for the one with a single node. Removing a node from any of the clusters will cause two clusters to have the same number of nodes. During a data collection process, the two clusters of the same size will have to do interleaving, which will not affect $t(N)_{\min}$. Therefore, the minimum number of time slots $t(N)_{\min}$ required by the BS is always equal to the number of clusters in the system. ∎

In contrast, for an MC2H network with N nodes organized in g clusters, where $N \geq g$, the number of time slots required by the BS to collect data from all sensor nodes is maximized when $N - (g - 1)$ nodes are belonging to the same cluster. The remaining $g - 1$ clusters will all have a cluster size of 1, and we have

$$t(N)_{\max} = \begin{cases} N - (g-1), & N > 2(g-1) \\ N - (g-1) + 1, & N = 2(g-1) \\ g, & \text{otherwise.} \end{cases} \tag{7.21}$$

In an SC network, the number of time slots required by the BS to collect data from all sensor nodes is minimized when the CH is at the middle of the chain, i.e.,

$$t(N)_{\min} = \begin{cases} \dfrac{N+1}{2} + 1, & N \text{ is odd} \\ \dfrac{N}{2}, & N \text{ is even,} \end{cases} \tag{7.22}$$

where N is the number of nodes in the network. In contrast, in an SC network, the number of time slots required by the BS to collect data from all sensor nodes is maximized when the CH is at the end of the chain, i.e.,

$$t(N)_{\max} = N, \tag{7.23}$$

where N is the number of nodes in the network.

In networks using MST and CTP, the number of time slots required by the BS to collect data from all sensor nodes is lower-bounded by Equations (7.12) and (7.13). On the other hand, the number of time slots required by the BS to collect data from all sensor nodes is maximized when the resultant networks of MST and CTP are in a single-cluster two-hop structure, which is upper-bounded by $t(N)_{max} = N$.

7.6   SIMULATIONS

In this section, the proposed network structure is compared with an MC2H network, an SC network, an MST network, and a CTP network. Networks having N nodes—with N varying from 4 to 64, with a step size of 4, were distributed randomly and evenly on a sensing field of 50×50 m^2. The center of the sensing field was located at $(x,y) = (25$ m, 25 m$)$. In the simulations, synchronization among wireless sensor nodes was maintained by the physical layer and the data link layer. Wireless sensor nodes were assumed to be equipped with CDMA (code division multiple access)-based transceivers [15]. Interference due to parallel transmissions can be alleviated by utilizing different spreading sequences in different data links. Media access control during network formation was handled by the MAC (media access control) sub-layer and was assumed to be satisfactory. A node can either receive or transmit at any time. In the simulation, a wireless sensor node was always capable of fusing all received packets into a single packet by means of data/decision-fusion techniques, and the size of an aggregated packet was independent of the number of packets received. For each network, the averaged data collection time (DCT) was used to indicate its data collection efficiency. The communication distance of a network was represented by the following function:

$$\psi = \sum_{i=1}^{N} u_i d_{i\_B}^h + \sum_{i=1}^{N-1} \sum_{j=i+1}^{N} c_{ij} d_{ij}^h, \tag{7.24}$$

where u_i is a value that indicates the number of CHs ($u_i = 1$) and CMs ($u_i = 0$). Parameter d_{i_B} is the distance between a CH and the BS. Here, c_{ij} is an indicator to indicate the presence ($c_{ij} = 1$) or absence ($c_{ij} = 0$) of a data link between node i and node j. Furthermore, d_{ij} is the geographical distance between nodes i and j. In the simulations, the path-loss exponent h was assumed to be 2. The BS was assumed to be at the center of the sensing field (i.e., $x = 25$ m, $y = 25$ m).

For the simulations on network lifetime, each node was given 50 J of energy. The energy model of the wireless sensor nodes was the same as the one introduced in Section 7.4. A network periodically performed the data collection process. The lifetime of a network was defined as the number of data collection processes (in terms of rounds) that a network can accomplish before any of its nodes runs out of energy. Each data packet was p_{data} bits long. Other packets were all regarded as control packets. Each control packet was p_{ctrl} bits long. Values of the parameters used in the energy model are shown in Table 7.2. The network structures under test were classified into two types: Type I, a single-cluster network structure; and Type II, a multiple-cluster network structure. Under this classification, all of the tested structures belong to Type I, whereas only MC2H and the proposed network structure belongs to both types. An MC2H network becomes a single cluster two-hop network when it is implemented as a type II network. Note that the number of clusters in

TABLE 7.2

Values of Parameters Used in the Simulations

Parameters	Values
$E_{i\_TCR\_RX}$	50×10^{-6} J/bit
$E_{i\_TCR\_EC}$	50×10^{-6} J/bit
$E_{i\_TCR\_PA}$	100×10^{-9} J/bit/m^2
E_{i_MCU}	5×10^{-6} J/bit
E_{i_SB}	50×10^{-6} J/bit
p_{data}	1024 bits
p_{ctrl}	64 bits

Type I and Type II network structures are different, and thus the results obtained by different types of network structures should not be compared directly.

For the proposed network structure to work as a Type I structure, sufficient dummy nodes had to be added. To work as a Type II structure, the proposed network structure could be constructed without adding any dummy nodes. The cluster number of the MC2H network was fixed to 1 when it was worked as a Type I structure. Here, ψ of the MC2H network was minimized by selecting the node with minimum separation from its fellow nodes as the CH. To work as a Type II structure, CHs in the MC2H networks were selected in a random manner as given in the literature [2], while the optimum number of CHs was selected according to Equation (7.20). In both configurations, CMs were connected to their nearest CH.

The SC network can only work as a Type I structure, and the chain is formed by using a greedy algorithm, as given by Lindsey and Raghavendra [3]. To minimize DCT, the node closest to the middle of the chain (in terms of hops) was selected as the CH. Similar to the SC network, the MST network can only work as a Type I structure. Networks were formed by using Prim's algorithm as given by Tan and Körpeoğlu [4]. To minimize DCT, the node with the smallest separation (in terms of hops) to all leaf nodes was selected as the CH. In networks using CTP, the node closest to the center of the sensing terrain was regarded as the root of the collection tree. The ETX of a path was expressed as the squared value of the path length [6]. The root of the tree had an ETX of 0, where the ETX of an arbitrary node was its cumulated ETX, through its parent nodes, to the root [5]. Each node chose its best route by selecting the path with the minimum cumulated ETX. Simulation results are shown in Figures 7.3–7.8. All results presented in this chapter were averaged over 100 simulations.

7.7 ANALYSIS

As expected from the discussion in Section 7.5, the DCT of networks with the proposed network structure was the lowest among the Type I structures. In simulations among the five Type I structures, the DCT of networks with the proposed network structure was the lowest, followed by networks with CTP. Since the aim of the MST

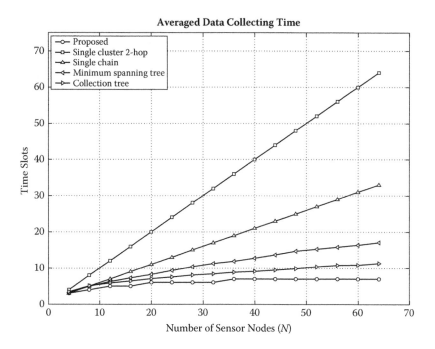

FIGURE 7.3 Averaged data collection time of different single-tree structures.

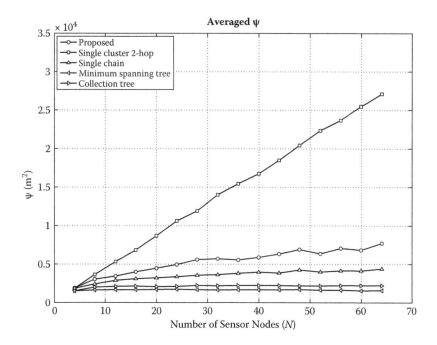

FIGURE 7.4 Averaged ψ of different single-tree structures.

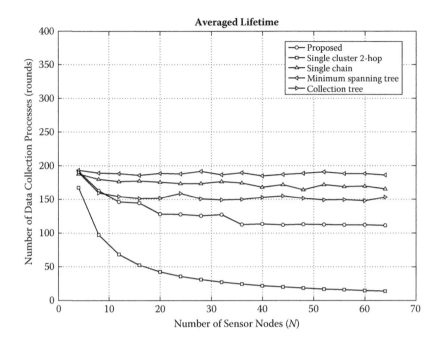

FIGURE 7.5   Averaged lifetime of different single-tree structures.

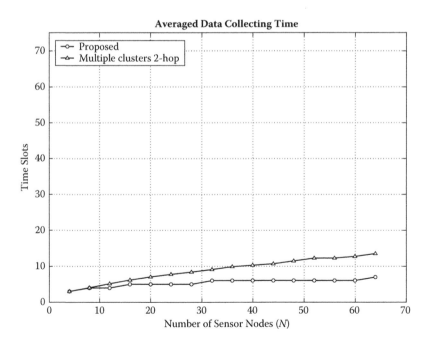

FIGURE 7.6   Averaged data collection time of different multiple-cluster structures.

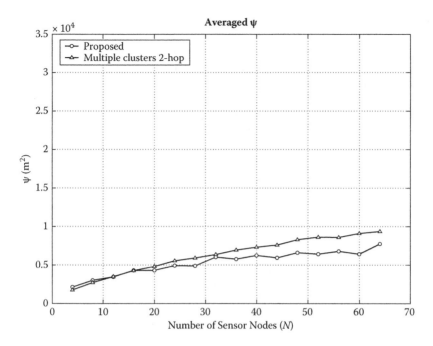

FIGURE 7.7 Averaged ψ of different multiple-cluster structures.

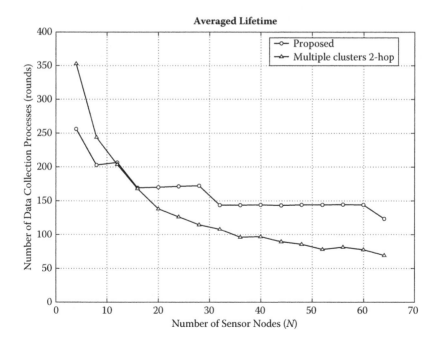

FIGURE 7.8 Averaged lifetime of different multiple-cluster structures.

is to minimize the total weight of edges, it did not perform well in reducing bottle-necks, and therefore it ranked fourth. In an SC network, it takes a very long time for data to propagate from both ends of the chain to the CH at the middle. This explains why SC networks have much higher DCT than networks with the proposed network structure. With the two-hop (SC2H) structure, the networks with MC2H structure did not have any advantage in reducing DCT. The MC2H network was the one with the highest DCT among Type I structures.

In terms of minimizing ψ, the MST network clearly ranked first. The ETX used in networks with CTP can greatly reduce the communication distances among sensor nodes, which explains its rank of second. Nodes in networks with SC structure try to reduce the total communication distance by connecting only to their nearest neighbors. This strategy is effective for networks with a small number of nodes. However, to maintain a single-chain structure, it is unavoidable for the SC network to increase its ψ as the number of nodes increases. The SC network therefore ranked third. Using the optimization techniques employed in Section 7.4, the ψ of networks constructed using the proposed algorithm did not increase drastically with increasing N, and this approach ranked fourth. With all sensor nodes connected to a single CH, the MC2H network was unquestionably the structure with the highest ψ.

In a data collection process, a node with connection degree k is required to receive $k - 1$ data packets, perform $k - 1$ times of data fusion, and transmit a single data packet. A node with a high connection degree will certainly consume more energy that one with a low connection degree. Therefore, a network that has a uniform-connection degree distribution is more likely to yield a longer network lifetime than one with a nonuniform-connection degree distribution. The minimum-spanning property of an MST network can make the connection degree evenly distributed among nodes. Therefore, the MST network can achieve the highest network lifetime for networks with $N > 4$. In an SC network, where most of the nodes have connection degrees equal to 2, the energy consumed by each node in receiving and fusing data is relatively low. Therefore, the SC network ranked second. Similar to the simulations on the ψ, the network lifetime of networks with CTP and the proposed network structure ranked third and fourth, respectively. In an MC2H network, all CMs are connected to a single CH. The CH is heavily loaded and has a very high energy consumption, which explains why an MC2H network had the lowest network lifetime among all of the tested network structures.

In simulations of the two Type II structures (the proposed structure and MC2H), networks formed by the proposed algorithm were shown to have the lowest DCT. Although the networks with MC2H structure had been tuned to give the optimum number of clusters, there was no control on the distribution of sensor nodes in each cluster. The DCT of networks with MC2H structure therefore greatly increased as N increased. For the same reason, the network lifetime of networks with MC2H structure decreased gradually as N increased. In terms of ψ, both the proposed and the MC2H network structures gave similar results when $N \leq 12$. For $N > 12$, networks constructed by the proposed algorithm had lower ψ than those constructed in the MC2H structure. The gap increased further with increasing N. According to Equation (7.20), an MC2H network is most efficient if there are g_{opt} clusters, where g_{opt} is proportional to N. As N increases, g_{opt} increases, and thus more nodes are involved in long-distance transmissions. The same thing happens to networks constructed by the proposed algorithm. Nevertheless, due to the special topology of the proposed network structure, the number of clusters increased at a lower rate. This explains why the ψ of networks formed by the proposed algorithm was lower than the ψ constructed in the MC2H structure. This also explains why networks formed by the proposed algorithm can obtain a longer network lifetime than those constructed with the MC2H structure.

7.8 CONCLUSIONS

In this chapter, a delay-aware data collection network structure and its formation algorithm are proposed. To cater for different applications, the network-formation algorithm can be implemented in either a centralized or decentralized manner. The performance of the proposed network structure is compared with a multiple-cluster two-hop network structure, a single-chain network structure, a minimum-spanning tree network structure, and a collection tree network structure. The proposed network structure is shown to be the most efficient in terms of data collection time among all of these network structures. The proposed network structure can greatly reduce the data collection time while keeping the total communication distance and the network lifetime at acceptable values.

REFERENCES

1. Al-karaki, J. N., and A. E. Kamal. 2004. Routing techniques in wireless sensor networks: A survey. *IEEE Wireless Communications Mag.* 11 (6): 6–28.
2. Heinzelman, W. B., A. P. Chandrakasan, and H. Balakrishnan. 2002. An application-specific protocol architecture for wireless microsensor networks. *IEEE Trans. Wireless Communications* 1 (4): 660–70.
3. Lindsey, S., and C. S. Raghavendra. 2002. PEGASIS: Power-efficient gathering in sensor information systems. In *Proceedings of the Aerospace Conference*, vol. 3, 1125–30. Piscataway, NJ: IEEE Press.
4. Tan, H. Ö., and Í. Körpeoğlu. 2003. Power efficient data gathering and aggregation in wireless sensor networks. *ACM SIGMOD Record* 32 (4): 66–71.
5. Fonseca, R., O. Gnawali, K. Jamieson, S. Kim, P. Levis, and A. Woo. 2007. The collection tree protocol. *TinyOS Enhancement Proposals (TEP)* 123 (December).
6. Couto, D. S. J. D. 2004. High-throughput routing for multi-hop wireless networks. PhD diss., Massachusetts Institute of Technology, Dept. Electrical Engineering and Computer Science.
7. Tekdas, O., J. H. Lim, A. Terzis, and V. Isler. 2009. Using mobile robots to harvest data from sensor fields. *IEEE Wireless Communications Mag.* 16 (1): 22–28.
8. Manjeshwar, A., and D. P. Agrawal. 2001. TEEN: A routing protocol for enhanced efficiency in wireless sensor networks. In *Proceedings of 15th International Parallel and Distributed Processing Symposium (IPDPS)*, 2009–15. Piscataway, NJ: IEEE Press.
9. Manjeshwar, A., and D. P. Agrawal. 2002. APTEEN: A hybrid protocol for efficient routing and comprehensive information retrieval in wireless sensor networks., In *Proceedings of 16th International Parallel and Distributed Processing Symposium (IPDPS)*, 195–202. Piscataway, NJ: IEEE Press.

10. Florens, C., M. Franceschetti, and R. J. McEliece. 2004. Lower bounds on data collection time in sensory networks. *IEEE Jour. Selected Areas in Communications* 22 (6): 1110–20.

11. Wang, W., Y. Wang, X.-Y. Li, W.-Z. Song, and O. Frieder. 2006. Efficient interference-aware TDMA link scheduling for static wireless networks. In *Proceedings of 12th Annual International Conference on Mobile Computing and Networking (MobiCom)*, 262–73. New York: ACM Press.

12. Solis, I., and K. Obraczka. 2004. The impact of timing in data aggregation for sensor networks. In *Proceedings of International Conference on Communications*, vol. 6, 3640–45. Piscataway, NJ: IEEE Press.

13. Chen, Z. Y., and X. F. Wang. 2006. Effects of network structure and routing strategy on network capacity. *Phys. Rev. E* 73 (3): 036107.

14. Song, M., and B. He. 2007. Capacity analysis for flat and clustered wireless sensor networks. In *Proceedings of International Conference on Wireless Algorithms, Systems and Applications (WASA)*, 249–53. Piscataway, NJ: IEEE Press.

15. Texas Instruments. 2007. 2.4 GHz IEEE 802.15.4/Zigbee® RF transceiver. Datasheet CC2520. http://focus.ti.com/lit/ds/symlink/cc2520.pdf.

8 Distributed Modulation Classification in the Context of Wireless Sensor Networks

Jefferson L. Xu, Wei Su, and Mengchu Zhou

CONTENTS

8.1 INTRODUCTION

Automatic modulation classification (AMC) is a desirable technique for cognitive radios in detecting and identifying the primary use of the dynamic spectrum access, in classifying the modulation scheme of an adaptive modulation transmission, and in estimating the physical layer parameters of an ad hoc communication network.

The AMC technique has been widely used in both military and commercial applications [1]. It is a statistical signal-processing method to recognize the modulation scheme of a noisy communication signal based on multiple hypotheses. It is generally developed using feature-based or likelihood ratio test (LRT)-based algorithms.

The former employs characteristics such as moments [2] or cumulants [3, 4] and makes decisions based on the best estimates of desired features. The LRT-based solutions [5, 6] compute the likelihood functions of the received signal under different hypotheses and compare the likelihood ratio against a Bayesian-criterion-determined threshold to make judgments. Comparing to LRT-based approaches, feature-based methods in essence have more limitations on suitable modulation schemes. Furthermore, their performance upper bounds are quantitatively lower than those of LRT-based methods, according to several studies in the literature [7, 8].

Over the last two decades, extensive research has been conducted on AMC methods, but most of them have used a single communication receiving sensor. The performance presented in terms of successful or false classification rate largely depends on the channel quality and the signal strength on the receiver side. However, in many noncooperative communication scenarios, the channel is unfavorably bad and the observed signal is usually weak. In cognitive radio network applications, a radio is able to minimize the transmission power based on the channel condition, or it may hop through multiple adjacent relay nodes with lower transmitting power. Novel technology is desirable for sensing and classifying weak signals reliably in a noncooperative and dynamic communication environment.

With the emergence of software-defined radios and wireless radio networks in modern communication systems, distributed signal sensing and classification shows promise as a means of achieving revolutionary AMC performance. Software-defined radios may be programmed to perform the distributed signal sensing and modulation classification, and local classification decisions will be collected by the decision maker via the communication network. Techniques have been developed to eliminate the signal distortion introduced by the channel. Assuming the channel parameters are independent random variables, combined signals from a set of geographically dispersed sensors may provide a better statistical description than any individual node alone. Therefore, the problem of AMC is expected to be more reliably and successfully resolved if more than one receiver participates in the classification. Distributed signal fusion, also known as centralized fusion, relays the distributed signal snapshots to the center node, and the decision maker fuses the distributed snapshots in the central node for the global decision. Precision timing, frequency, and phase synchronization will be needed. While the distributed-decision fusion makes hard or soft decisions locally, the decision maker in the central node fuses the local decision to form the global decision. This eliminates the need for synchronization and high transmission bandwidth. The pros and cons between the two fusion methods have been discussed by Su and Kosinski [6].

This chapter focuses on the distributed-decision fusion method by studying an optimized solution for distributed modulation classification. LRT-based AMC approaches are adopted in this chapter, but almost all known AMC algorithms can be extended to distributed AMC framework. In Section 8.2, we briefly review the LRT-based modulation classification and formulate its performance evaluation method. Section 8.3 introduces a new method of LRT-based modulation classification, which can work well in the context of sensor networks under the constraint on the communication bandwidth. The new classification procedure is explained in detail, and numerical results are presented as well to show the performance enhancements by sensor networks with different structures. Finally, Section 8.4 provides some concluding remarks.

8.2 SINGLE-RECEIVER MODULATION CLASSIFICATION

8.2.1 The Construction of Likelihood Function

Consider the problem of identifying the modulation type of the message sequence $s(t)$ that propagates over a physical transmission medium and encounters an additive noise $n(t)$. As the data source of decision theory, the noiseless digital modulated signal is conditionally deterministic. It needs to be modeled via a comprehensive form as follows:

$$s(t, \mathbf{u}_i) = \sum_{n=0}^{N-1} s_n^{k,i} e^{j(\omega_c t + \theta_c)} g(t - nT_s - \varepsilon T_s), \tag{8.1}$$

where $\mathbf{u}_i$ is defined as a multiple dimensional parameter space in which a set of signal and channel parameters are stochastic or deterministic unknown variables under the hypothesis, H_i, that the ith type of modulation signal is being received.

$$\mathbf{u}_i = \left\{ \theta_c, \varepsilon, h(t), \left\{ s_n^{k,i} \right\}_{k=1}^{M_i}, \omega_c \right\}. \tag{8.2}$$

The notations used in Equations (8.1) and (8.2) are

N is the number of symbols inside each received data frame.

$g(t)$ is the composite effect of the residual channel denoted as $h(t)$ and the standard pulse-shaping function.

$\{s_n^{k,i}\}$ is the alphabet, i.e., the vector of complex symbols belonging to the ith modulation scheme that has M_i known constellation symbols, $k \in \{1, 2, \ldots, M_i\}$, and $1 < i \le \xi$, ξ is the cardinality of the constellation symbol set. The objective of modulation classification is to recognize the constellation from ξ different candidates. During the nth time interval, one symbol from the alphabet, denoted by $s_n^{k,i}$, is transmitted. The average power of each symbol is usually normalized to unit 1.

θ_c is the carrier-phase offset introduced by the propagation delay as well as the initial carrier phase on the transmitter side. θ_c is normally assumed to vary slowly and be consistent over a period of observation intervals.

ε is the normalized epoch for timing offset between the transmitter and the receiver, $0 \le \varepsilon < 1$.

ω_c is the residual carrier frequency after frequency down-conversion and coarse carrier removal.

To solve the problem in the classical detection theory, the incoming signal is expressed in terms of a vector $\mathbf{R}$ that consists of a set of N noisy complex observations: $\mathbf{R} = [r_0, r_1, \cdots, r_{N-1}]^T$. The complex instant r_n is the discrete output of the pulse-shape-matched filter, and the superscript T is the transpose operator. Consider that the channel noise between r_n and the emitted symbol is assumed to be the complex envelope of real white Gaussian noise with power spectrum density (PSD) $N_0/2$. The real and imaginary parts of the noise complex envelope are independent and identically distributed (i.i.d.) with variance $\sigma^2 = N_0/2$. The parameterized probability density function (PDF) of r_n under hypothesis H_i is explicitly written as

$$p(r_n|H_i,\mathbf{u}_i) = \frac{1}{2\pi\sigma^2}\exp\left(\frac{-|r_n - s_n(\mathbf{u}_i)|^2}{2\sigma^2}\right), \tag{8.3}$$

where $s_n(\mathbf{u}_i)$ is the average of $s(t,\mathbf{u}_i)$ over the time interval between nT_s and $(n + 1)T_s$. This would be valid when all parameters in $\mathbf{u}_i$ vary slowly enough and the epoch $\varepsilon = 0$. In statistics, $p(r_n|H_i,\mathbf{u}_i)$ is denoted and viewed as the likelihood function of H_i given each single observation, r_n. In other words, it indicates how likely the ith modulation is in light of the incoming signals. The first term in Equation (8.3) is a fixed value, and the variable $|r_n|^2$ in the exponent is independent of various hypotheses. Ignoring them without a loss in generality, and replacing the noise variance with $N_0/2$, the likelihood function of a parameterized model is expressed as

$$\Gamma(r_n \mid H_i,\mathbf{u}_i) = \exp\left(-\frac{|(s_n\mathbf{u}_i)|^2 - 2\,\mathrm{Re}\{r_n^* s_n(\mathbf{u}_i)\}}{N_0}\right), \tag{8.4}$$

For the remainder of this chapter, the expression in Equation (8.4) acts as the baseline for the likelihood function in modulation classification. It also governs the mapping from the parameter space $\mathbf{u}_i$ to the observation space based on the composite hypothesis H_i that is characterized by $\mathbf{u}_i$. If the signal and channel parameters inside $\mathbf{u}_i$ are treated as random variables with known density functions, i.e., if the statistical characteristic of $\mathbf{u}_i$ is known, $p(r_n|H_i,\mathbf{u}_i)$ can be designated as the joint PDF of the random variables in $\mathbf{u}_i$. To maximize the likelihood under each hypothesis, the marginalized density of a single observation takes the statistical averaging of Equation (8.4):

$$\Gamma(r_n|H_i) = \mathrm{E}_{\mathbf{u}_i}[\Gamma(r_n|H_i,\mathbf{u}_i)], \tag{8.5}$$

where $\mathrm{E}[\cdot]$ is the expectation operator.

In this chapter, we consider the ambiguity of the emitted symbol and the carrier-phase offset, which might be induced by an imprecise carrier recovery. In the absence of sufficient carrier-phase information, θ_c is assumed to vary uniformly over the interval $[-\pi, \pi]$. Normally, $s^{k,i}$ is modeled as a discrete random variable uniformly distributed over the alphabet set. The calculation of the likelihood function of r_n is explicitly written as:

$$\Gamma(r_n|H_i) = \int_{-\pi}^{\pi}\frac{1}{2\pi M_i}\sum_{k=1}^{M_i}\exp\left(-\frac{\left|s_n^{k,i}e^{j\theta_c}\right|^2 - 2Re\left\{r_n^* s_n^{k,i}e^{j\theta_c}\right\}}{N_0}\right)d\theta_c. \tag{8.6}$$

After collecting a sequence of N consecutive and independently transmitted symbols, $\mathbf{R} = [r_1, r_2, \ldots, r_N]^T$, we can calculate the composite likelihood function of the observation vector as follows:

$$\Gamma(\mathbf{R}|H_i) = \prod_{n=0}^{N-1} \Gamma(r_n|H_i). \tag{8.7}$$

Furthermore, because the natural logarithm is monotonically increasing and $\Gamma(\mathbf{R}|H_i)$ is positive, the log-likelihood function for the observation vector $\mathbf{R}$ is equivalently expressed as

$$\mathcal{L}(\mathbf{R}|H_i) = \ln\Gamma(\mathbf{R}|H_i). \tag{8.8}$$

The log-likelihood function reaches the maximum value at the same points as the likelihood function itself. The advantage of using the log-likelihood function will be discussed in the next subsection, where the quantified classification accuracy is derived.

8.2.2 MAXIMUM-LIKELIHOOD CRITERION

The goal is to find the decision rule that picks the most likely hypothesis and minimizes the average probability of error decisions. The use of Bayes's theorem offers the maximum-likelihood (ML) criterion as a statistical solution to the problem of classifying M-ary linear digital modulation schemes. Assume that the cost of the correct detection is zero and that the cost of a wrong detection is a constant. The optimum Bayesian criterion leads to a likelihood-ratio test. As a method of statistical hypothesis testing, LRT is widely applied to fault diagnoses [9], pattern recognition, and segmentation [10]. In a binary hypothesis detection problem, the criterion is constructed as the ratio of the function in Equation (8.7) under two hypotheses: H_i and H_j with prior probabilities P_i and P_j, respectively [11].

$$\frac{\Gamma(\mathbf{R}|H_i)}{\Gamma(\mathbf{R}|H_j)} \underset{H_j}{\overset{H_i}{\gtrless}} \eta = \frac{P_j}{P_i} \tag{8.9}$$

The test performed by Equations (8.6)–(8.9) is termed as the *average likelihood-ratio test* (ALRT) in consideration of the statistical averaging completed in Equation (8.5). Without any arithmetic approximation, the threshold η is unit 1 if all the modulations are equally likely a priori, i.e., $P_i = P_j$. Obviously, the binary hypothesis test can be generalized to ξ hypotheses with uniform prior probabilities. The detection is carried out according to

$$i_{max} = \underset{i \in \{1,2,\ldots,\xi\}}{\operatorname{argmax}} \Gamma(\mathbf{R}|H_i). \tag{8.10}$$

8.2.3 CLASSIFICATION PERFORMANCE

In this subsection, we attempt to quantify the performance of the modulation classification with the maximum-likelihood method. The evaluation of the classifier is measured by the probability of correct classification or by the probability of error decisions. In classifying ξ equiprobable modulation schemes, the performance of making correct decisions by certain rules is measured by the average probability of correct classification [12],

$$P_c = \frac{1}{\xi} \sum_{i=1}^{\xi} P_c(\mathbf{R}|H_i), \tag{8.11}$$

where $P_c(\mathbf{R}|H_i)$ denotes the correct probability associated with declaring that the ith modulation type is employed when it is the true type. The mathematical analysis of an ML-based classifier performance follows immediately from its criteria. In the case of binary classification, the conditional probabilities $P_c(\mathbf{R}|H_i)$ and $P_c(\mathbf{R}|H_j)$ are in principle calculated respectively by

$$P_c(\mathbf{R}|H_i) = \Pr\left(\frac{\Gamma(\mathbf{R}|H_i)}{\Gamma(\mathbf{R}|H_j)} \geq 1 | H_i\right) \tag{8.12}$$

and

$$P_c(\mathbf{R}|H_j) = \Pr\left(\frac{\Gamma(\mathbf{R}|H_i)}{\Gamma(\mathbf{R}|H_j)} < 1 | H_j\right). \tag{8.13}$$

It is more convenient to work in terms of a log-likelihood function with a collection of statistically independent observation vectors $\mathbf{R}$. The logarithm of the product $\Gamma(\mathbf{R}|H_i)$, defined as $\mathcal{L}(\mathbf{R}|H_i)$ in Equation (8.8), is the sum of the individual function $\mathcal{L}(r_n|H_i)$. The sum of N i.i.d. random variables $\mathcal{L}(r_n|H_i) - \mathcal{L}(r_n|H_j)$ satisfies the condition of the Central Limit Theorem. That is, $\mathcal{L}(\mathbf{R}|H_i) - \mathcal{L}(\mathbf{R}|H_j)$, which represents the difference between two log-likelihood functions under two different modulation hypotheses, approximates the normal distribution $\forall (N\mu_i, N\sigma_i^2)$, as N is sufficiently large. Thus Equation (8.12) becomes:

$$P_c(\mathbf{R}|H_i) = \int_{-\sqrt{N}\mu_i}^{\infty} \frac{1}{\sqrt{2\pi}\sigma_i} \exp\left(-\frac{\delta^2}{2\sigma_i^2}\right) d\delta, \tag{8.14}$$

where symbols μ_i and σ_i respectively denote the finite mean and variance of the variable $\mathcal{L}(\mathbf{R}|H_i) - \mathcal{L}(\mathbf{R}|H_j)$ under hypothesis H_i. The actual calculations can be performed numerically by using the following equations [13]:

$$\mu_i = \mathrm{E}\left[\mathcal{L}(r_n|H_i) - \mathcal{L}(r_n|H_j)|H_i\right] \tag{8.15}$$

and

$$\sigma_i^2 = E\left[\left(\mathcal{L}\left(r_n \mid H_i\right) - \mathcal{L}\left(r_n \mid H_j\right)\right)^2 \mid H_i\right] - \mu_i^2. \tag{8.16}$$

The expectations in Equations (8.15) and (8.16) are computed based on the PDF of r_n under the hypothesis H_i.

The lower integration limit inside Equation (8.14) has a key implication: The asymptotic behavior based on the number of processed symbols can be easily observed from it. The ML-based classifier is capable of identifying any constellation with zero error as the number of available observations, N, goes to infinity, even if the signal undergoes a very noisy transmission channel. A comparison of classification performance with different block length is illustrated in Figure 8.1. The detection is discriminating between modulation types PSK-16 and square shaped QAM-16. The probability curves calculated by Equations (8.11)–(8.16) are plotted as a function of the quality of the received signal, i.e., the signal-to-noise ratio (SNR) per symbol. Note that the curves with more observations horizontally move to the left relative to those obtained based on fewer symbols. The SNR difference between $N = 50$ and $N = 100$ is approximately 1 dB, and the difference between $N = 200$ and $N = 500$ is approximately 1.5 dB.

Unfortunately, processing an infinite number of received symbols is impractical in the real world. Alternatively, a parallel signal collection model may be considered to deal with the problem of modulation classification. In particular, there is a high probability that one channel is unreliable from time to time. In wireless communication, sensor networks are commonly used as a means of providing diversity against interference

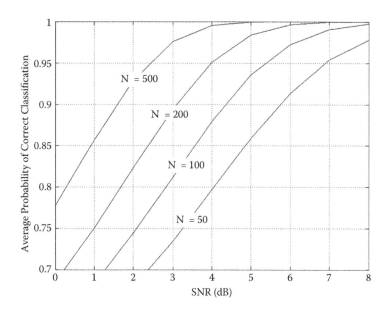

FIGURE 8.1 PSK-16 versus QAM-16 classification results using a different number of received symbols.

and channel fading. The next section discusses efforts to enhance the performance in modulation classification by having multiple receivers involved.

8.3 CLASSIFICATION VIA SENSOR NETWORKS

Prior sections are concerned with modulation classification for a single receiver. We now focus our attention on the problem of improving the accuracy of modulation classification via a wireless sensor network. We begin with the description of the sensor network model.

8.3.1 DISTRIBUTED MODULATION CLASSIFIER MODEL

A block diagram representation of this model is shown in Figure 8.2. A multisensor signal collection network is described by L communication sensors and a decision maker. The superscript ℓ is used to index the ℓth sensor, $\ell = 1, 2, \ldots,$ and L. An information-bearing signal sequence, $\{s_1, s_2, \ldots, s_N\}$ belonging to one of the modulation candidates, is broadcast by a transmitter and propagates over L parallel channels. The signal in each channel is corrupted by an additive zero-mean white Gaussian noise. The corruption processes among L diversity channels are assumed to be mutually statistically independent. Within the network, there are another L parallel independent channels that connect the final decision maker to these sensors, and are considered error free. Assuming that each sensor has the same number of observations, L noisy data sequences are collected simultaneously and processed noncooperatively by all sensors in the network. Due to the effect of different propagations and transmission environments, different signals are observed at receiving sensors even though the underlying base-band information sequences are identical.

If complete sensor observations can be nondestructively transmitted to the decision maker, a comprehensive optimization of hypothesis testing can be reached. But the constrained channel capacity within the sensor network makes the local raw observation sets inaccessible in their entirety at the decision maker. To relieve the requirements on the bandwidth of channels, we assume that local sensors have associated processors to have part of signal processing accomplished locally, such that

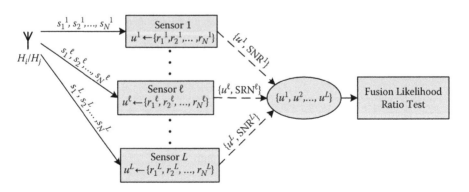

FIGURE 8.2   Two-stage modulation classification in the context of a sensor network.

the massive computations can be distributed to sensors. If we let each sensor perform LRT as is discussed in the prior section, there exist L preliminary binary decision results. Some research publications in the area of distributed-detection theory provide computationally difficult methods to determine local thresholds and obtain local and global optimality [14–20]. Person-by-person optimization is utilized in determining the threshold of the local LRT for each sensor. But in the realm of practical applications, identical sensors are always preferred from the perspectives of manufacture and deployment. To facilitate a more realistic detection, we set the local LRT threshold the same by constraining $\eta^\ell = P_j/P_i$ for all sensors in cases of binary-modulation-type classification. The local decision denoted by u^ℓ takes the value either i or j, and is transmitted to the decision maker without any interference for further processing. Finally, the decision maker determines which modulation scheme is more likely to be true after collecting all of the L local decisions. A demodulation method is chosen correspondingly from the demodulation toolbox based on the estimated modulation scheme, and the information-bearing signal is demodulated.

The design of a practical signal-fusion algorithm with the ability to improve the modulation classification accuracy is necessary for the distributed classification. The decision maker has no access to the raw input signal symbols, and the final decision is only affected by the preliminary decision vector, $\{u^1, u^2, ..., u^L\}$ produced based on the observations over N time instances. The observation period is relatively much longer than that of a single symbol. Hence, the strict synchronization among sensors is not required.

8.3.2 Detailed Derivations

A decision-maker-oriented composite modulation classification strategy that stems from the spirit of the Chair-Varshney decision fusion rule [16] is described here. From the standpoint of the decision maker, each u^ℓ is a binary random input variable that holds either i or j. Given the set of specified local decisions, $\{u^1, u^2, ..., u^L\}$, the optimal global modulation classifying strategy in a Bayesian sense can be obtained by an LRT conditioned on $\{u^1, u^2, ..., u^L\}$ [21]. Once again, the LRT performed in the decision maker for detecting the modulation scheme is constructed in cooperation with the threshold set to $\eta = P_j/P_i$

$$\frac{P(u^1, u^2, ..., u^L \mid H_i)}{P(u^1, u^2, ..., u^L \mid H_j)} \overset{H_i}{\underset{H_j}{\gtrless}} \eta. \qquad (8.17)$$

Different from the LRT performed based on the raw observations by local sensors, Equation (8.17) is used to find out which modulation type is more likely given the preliminary decisions made by other sensors. Two new definitions are generated after dividing these local decisions into two groups

$$\begin{cases} \mathbf{L}_i = \{\ell : u^\ell = i\} \\ \mathbf{L}_j = \{\ell : u^\ell = j\} \end{cases}. \qquad (8.18)$$

Because the local decisions are statistically independent, we have the conditional probability of the local decision set derived as follows:

$$P\left(u^1, u^2, \ldots, u^L | H_i\right) = \prod \Pr\left(u^\ell | H_i\right) = \prod_{\ell \in L_i} \Pr\left(u^\ell = i | H_i\right) \prod_{\ell \in L_j} \Pr\left(u^\ell = j | H_i\right). \quad (8.19)$$

Note that $\Pr(u^\ell = i | H_i)$ is a local version of probability of correct classification achieved by the ℓth sensor, and $\Pr(u^\ell = j | H_i)$ is the probability of making an erroneous decision. Complementing a superscript ℓ to the notations of conditional probabilities of correct classification, $P_c(\mathbf{R} | H_i)$, which was discussed in the prior section, the mapping to the probability in Equation (8.12) is described as

$$\begin{cases} \Pr(u^\ell = i | H_i) = P_c^\ell(\mathbf{R} | H_i) \\ \Pr(u^\ell = j | H_i) = 1 - P_c^\ell(\mathbf{R} | H_i) \end{cases} \quad (8.20)$$

Since an identical detection criterion is implemented in local sensors, the methods to obtain $P_c^\ell(\mathbf{R} | H_i)$ are the same across all sensors. But geographically dispersed sensors receive signals with various SNRs, which lead to different probabilities of correct classification. We use $P_c^{\ell,\text{SNR}}(\mathbf{R} | H_i)$ to represent achievable probabilities by the ℓth sensor under a certain SNR condition. Substituting Equation (8.20) into Equation (8.19), the fusion likelihood function under the ith hypothesis is characterized by

$$P\left(u^1, u^2, \ldots, u^L | H_i\right) = \prod_{\ell \in L_i} P_c^{\ell,\text{SNR}}\left(\mathbf{R} | H_i\right) \prod_{\ell \in L_j} \left(1 - P_c^{\ell,\text{SNR}}\left(\mathbf{R} | H_i\right)\right). \quad (8.21)$$

To the decision maker, the likelihood function under H_i certainly expects that more sensors within the network can produce a correct decision locally. Hence, more probabilities of a correct decision and fewer error probabilities are obtained. Similarly, the fusion likelihood function under H_j is

$$P\left(u^1, u^2, \ldots, u^L | H_j\right) = \prod_{\ell \in L_j} P_c^{\ell,\text{SNR}}\left(\mathbf{R} | H_j\right) \prod_{\ell \in L_i} \left(1 - P_c^{\ell,\text{SNR}}\left(\mathbf{R} | H_j\right)\right). \quad (8.22)$$

By combining Equations (8.21) and (8.22), we have the final detection criterion as follows:

$$\prod_{\ell \in L_i} \frac{P_c^{\ell,\text{SNR}}(\mathbf{R} | H_i)}{(1 - P_c^{\ell,\text{SNR}}(\mathbf{R} | H_j))} \prod_{\ell \in L_j} \frac{(1 - P_c^{\ell,\text{SNR}}(\mathbf{R} | H_i))}{P_c^{\ell,\text{SNR}}(\mathbf{R} | H_j)} \overset{H_j}{\underset{H_i}{\lessgtr}} \frac{P_j}{P_i}. \quad (8.23)$$

The probability, $P_c^{\ell,\text{SNR}}(\mathbf{R} | H_i)$, indeed reflects the accuracy of the classification algorithm performed at the ℓth local sensor. The ratio test in Equation (8.23) indicates that the individual preliminary result is weighted by the performance of the classifier itself and the signal strength or environmental condition of the observations. The distributed sensors equipped with a higher-accuracy classification algorithm or located in a friendly reception location make more contributions to the final score than others. The erroneous decision owing to a single "inaccurate" sensor becomes less likely with the alleviation by weighted average. Meanwhile, the increasing number of sensors selecting one certain hypothesis as the local decision makes the global decision more favorable to the corresponding modulation scheme.

8.3.3 IMPLEMENTATION

The implementation requires the knowledge of local sensor performance. The theoretical performance indices of LRT were introduced in Section 8.2.3. If the received signal condition SNR is available, theoretical values of $P_c^{\ell,\mathrm{SNR}}(\mathbf{R}|H_i)$ and $P_c^{\ell,\mathrm{SNR}}(\mathbf{R}|H_j)$ can be sufficiently well approximated with the aid of an asymptotic series. The intensive computation, however, is impossible to be accomplished on the fly. A more feasible implementation should perform the numerical calculation of both probabilities off-line. At the final decision maker node, the probabilities under all candidate modulation schemes are stored and indexed by SNRs with a proper resolution that is determined by the specific technique utilized to estimate SNR in each sensor.

The approach for distinguishing between two modulation types with a sensor network is summarized in the following five steps:

1. Each individual sensor performs preclassification after receiving a frame of N symbols. A local detection result is generated.
2. Each sensor keeps estimating the SNR of incoming signals.
3. The real-time estimated SNR is quantized and sent to the primary node along with the binary local detection result in each N symbol interval.
4. The theoretically correct probabilities $P_c^{\ell,\mathrm{SNR}}(\mathbf{R}|H_i)$ and $P_c^{\ell,\mathrm{SNR}}(\mathbf{R}|H_j)$ corresponding to the currently estimated SNR informed by the ℓth sensor are extracted from the storage memory.
5. The fusion LRT is executed using Equation (8.23).

8.3.4 PERFORMANCE EVALUATION AND SIMULATION RESULTS

In the following discussion, the performance of the distributed LRT-based modulation classifier is demonstrated under different network structures. The implemented local processing is common to all local sensors, while the condition of the signals received by each sensor may be different. The number of received signal symbols, N, in a single frame waiting to be processed is fixed to 100.

8.3.4.1 The Effect of the Number of Sensors

The distributed modulation classifier is first evaluated in the framework comprising different numbers of sensors, and the results are plotted in Figures 8.3–8.6. The classification is carried out to discriminate between two modulation hypotheses chosen

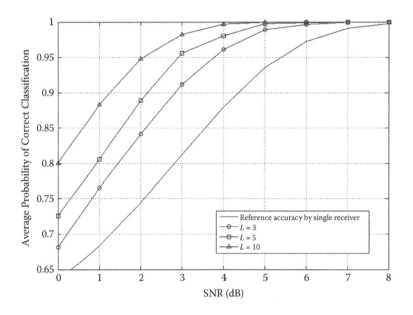

FIGURE 8.3 Comparison of the average probability of correct classification for sensor networks with $L = 3$, 5, and 10 sensors, PSK-16 versus QAM-16, $N = 100$.

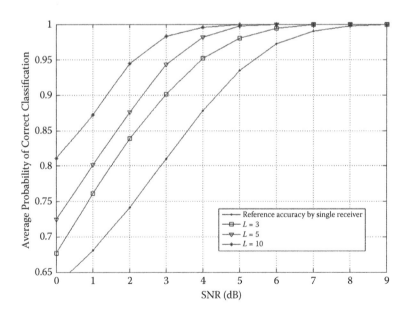

FIGURE 8.4 Comparison of the average probability of correct classification for sensor networks with $L = 3$, 5, and 10 sensors, QAM-32 versus PSK-8, $N = 100$.

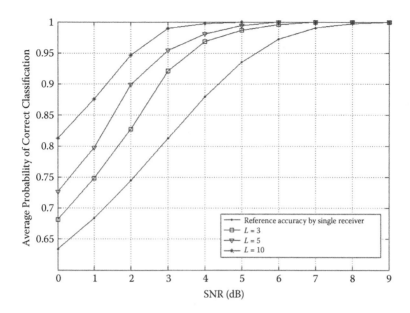

FIGURE 8.5 Comparison of the average probability of correct classification for sensor networks with $L = 3$, 5, and 10 sensors, PSK-8 versus QAM-16, $N = 100$.

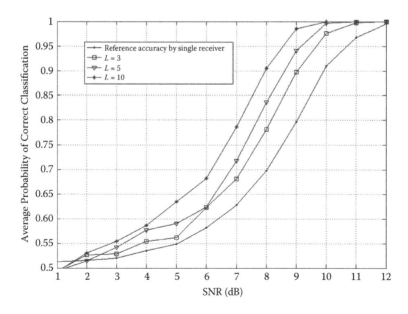

FIGURE 8.6 Comparison of the average probability of correct classification for sensor networks with $L = 3$, 5, and 10 sensors, PSK-8 versus PSK-16, $N = 100$.

from the set {PSK-8, PSK-16, QAM-16, QAM-32} in which each constellation is equally likely to be transmitted. The carrier-phase offset is random to all of the sensors. Each emitted symbol is corrupted by white noise and rotated by a randomly generated carrier-phase offset that is a constant value across 100 symbols.

The local preliminary modulation classifications are first executed under the assumption that all of the local observations have the same SNR. The fusion LRT is performed with Equation (8.23). The overall classification performance is represented in terms of average probability of correct classification measured from 1000 Monte Carlo simulations for each modulation type against various SNR levels. The theoretical counterpart derived for a single receiver is also plotted as a reference.

In comparison with the detection performed by a single receiving sensor, the curves in Figures 8.3–8.6 exhibit 1.5-dB, 2-dB, and 3-dB performance gains in SNR as the number of sensors, L, increases to 3, 5, and 10, respectively. In the case of PSK-8 versus PSK-16 using the same number of sensors, the performance enhancement is 1 dB, 1.5 dB, and 2 dB, respectively. Clearly, the more sensors, the better is the classification accuracy.

8.3.4.2   The Network with Sensor Output Misclassifications

Due to the transmission distance, fading, and geographical dispersion, the channels from the source to the sensor are in practice different from each other. Some sensors can produce much higher misclassification probabilities than others. As a participant in a composite decision algorithm, each single sensor should influence the final classification accuracy positively, or at least equally, under any circumstance. Even if the quality of the observed signals is absolutely low, the high possibility of wrong local decisions produced by the sensor should not degrade the overall performance after the fusion of local decisions. Putting this basic expectation in another way, each new sensor should not be added to the network in a harmful manner.

Next, we address this expectation to the distributed classifier by creating an experiment to simulate the scenario where the sensor network consists of identical sensors, but the transmission channels between the transmitter and sensors are not identical. In the worst-case scenario where there is only one sensor, the dark node shown in Figure 8.7 is capable of receiving signals as expected, and the signal strength received by this sensor and the noise floor of its surroundings construct experimental SNR values. The rest of the network is intentionally deployed in an extremely weak reception environment, or all of the other sensors are supposed to concurrently be subject to a large noise. The signal received by other sensors always has SNR = 0 dB. The reason for keeping one sensor working properly is to obtain a comparison of the performance with a single receiver.

Distinguishing between PSK-16 and QAM-16 is repeated under the same conditions that the simulation results in Figure 8.3 were obtained. The resulting curves are plotted in Figure 8.8. The obtained average probabilities with 3, 5, and 10 sensors still outperform those with a single receiver when only one sensor receives experimental signals. When all of the sensors in the network receive 0 dB SNR signals, the classification accuracies are the same as the ones shown in Figure 8.3. Thus, the primary node with new fusion LRT bases its decision on the sensor with higher

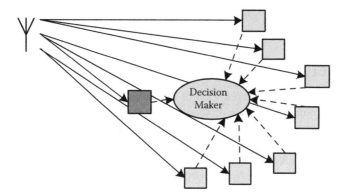

FIGURE 8.7 The experiment to evaluate the effect of multiple sensors in the network, which produce high probabilities of local error classifications.

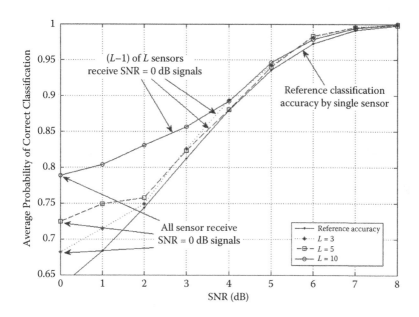

FIGURE 8.8 Performance curves for the sensor networks with nonidentical channels between the transmitter and sensors. The sensor network has only one sensor receive signal with SNR per symbol shown on the x-axis; $L = 3$, 5, and 10 sensors, PSK-16 versus QAM-16, $N = 100$.

detection accuracy. Other sensors are proven not to induce any negative contribution to the classification performance when they are located in a hostile environment.

8.4   CONCLUDING REMARKS

This chapter explored the application of the distributed-decision fusion rule to the problem of automatic modulation classification. An LRT-based modulation classification algorithm is pre-executed in local sensors. On the basis of preliminary local results, a second fusion likelihood-ratio test is executed by the decision maker. Due to the slow data rate required within the network, the synchronization among sensors is not an issue in this approach. Performance evaluations reveal that the identification reliability is improved consistently as more distributed receivers participate in the classification. The sensors will not induce a potential negative contribution to the classification performance when they are located in a hostile environment.

For the simplicity of discussion and mathematical analysis, the detection method presented in this chapter is confined to the classification between two modulation candidates. The generalization to multiple modulation types can be achieved by extending Equation (8.19) to multiple hypotheses. In a work by Zhang, Ansari, and Su [22], the multiple local decisions are discussed so that more than two types of modulations can be classified. A precalculated confusion table is used to provide the probabilities of classification to the decision maker for estimating the global decision. In a work by Xu, Su, and Zhou [23], the local sensor is no longer responsible for making any local hard decision. Instead of the local classification result, the information provided to the decision maker is the log-likelihood function values that are computed based on the local incoming symbols without contamination of local processing. In other words, the distributed sensors make only soft decisions by sending scores of local classification results to the decision maker. The decision maker consolidates all scores to calculate the global decision. The soft-decision method has demonstrated more reliable results than the hard-decision method.

ACKNOWLEDGMENT

The third author was in part supported by the National Basic Research Program of China under Grant No 2011CB302804.

REFERENCES

1. Su, W., J. Kosinski, and M. Yu. 2006. Dual-use of modulation recognition techniques for digital communication signals. In *Proceedings of Systems, Applications and Technology (LISAT)*, 1–6. Piscataway, NJ: IEEE Press.
2. Soliman, S. S., and S. Z. Hsue. 1992. Signal classification using statistical moments. *IEEE Trans. Commun.* 40 (5): 908–16.
3. Swami, A., and B. M. Sadler. 2000. Hierarchical digital modulation classification using cumulants. *IEEE Trans. Commun.* 48 (3): 416–29.
4. Dobre, A., Y. Bar-Ness, and W. Su. 2003. Higher-order cyclic cumulants for high order modulation classification. In *Proceedings of Conference on Military Communications (MILCOM)*, vol. 1, 112–17. Washington, DC: IEEE Computer Society.

5. Hameed, F., O. A. Dobre, and D. C. Popescu. 2009. On the likelihood-based approach to modulation classification. *IEEE Trans. Wireless Commun.* 8 (12): 5884–92.

6. Su, W., and J. Kosinski. 2003. Comparison and simulation of digital modulation recognition algorithms. In *International Symposium on Advanced Radio Technologies 2003, ISART 2003 IEEE*, Boulder, CO, Mar.

7. Dobre, O. A., and H. Fahed. 2006. Likelihood-based algorithms for linear digital modulation classification in fading channels. In *Proceedings of Canadian Conference on Electrical and Computer Engineering (CCECE)*, 1347–50. Piscataway, NJ: IEEE Press.

8. Dobre, O. A., A. Abdi, Y. Bar-Ness, and W. Su. 2007. Survey of automatic modulation classification techniques: Classical approaches and new trends. *IET Commun.* 1:137–56.

9. Li, P., and V. Kadirkamanathan. 2001. Particle filtering based likelihood ratio approach to fault diagnosis in nonlinear stochastic systems. *IEEE Trans. Systems, Man, and Cybernetics, Part C: Applications and Reviews* 31 (3): 337–43.

10. Cohen, F. S., Z. Fan, and S. Attali. 1991. Automated inspection of textile fabrics using textural models. *IEEE Trans. Pattern Analysis and Machine Intelligence* 13 (8): 803–8.

11. Trees, H. L. V. 2001. *Detection, estimation, and modulation theory*, Part 1. New York: John Wiley & Sons.

12. Sills, J. A. 1999. Maximum-likelihood modulation classification for PSK/QAM. In *Proceedings of Military Communications Conference (MILCOM)*, vol. 1, 217–20. Piscataway, NJ: IEEE Press.

13. Wei, W., and J. M. Mendel. 2000. Maximum-likelihood classification for digital amplitude-phase modulations. *IEEE Trans. Commun.* 48:189–93.

14. Tenney, R. R., and N. R. Sandell. 1981. Detection with distributed sensors. *IEEE Trans. Aerospace and Electronic Systems* 17 (4): 501–10.

15. Reibman, A. R., and L. W. Nolte. 1987. Optimal detection and performance of distributed sensor systems. *IEEE Trans. Aerospace and Electronic Systems* 23 (1): 24–30.

16. Chair, Z., and P. K. Varshney. 1986. Optimal data fusion in multiple sensor detection systems. *IEEE Trans. Aerospace and Electronic Systems* 22 (1): 98–101.

17. Willett, P., P. F. Swaszek, and R. S. Blum. 2000. The good, bad and ugly: Distributed detection of a known signal in dependent Gaussian noise. *IEEE Trans. Signal Processing* 48 (12): 3266–79.

18. Chen, B., L. Tong, and P. K. Varshney. 2006. Channel-aware distributed detection in wireless sensor networks. *Signal Processing Magazine, IEEE* 23 (4): 16–26.

19. Duman, T. M., and M. Salehi. 1998. Decentralized detection over multiple-access channels. *IEEE Trans. Aerospace and Electronic Systems* 34 (2): 469–76.

20. Hoballah, I. Y., and P. K. Varshney. 1989. Distributed Bayesian signal detection. *IEEE Trans. Information Theory* 35 (5): 995–1000.

21. Liu, B., and B. Chen. 2006. Channel-optimized quantizers for decentralized detection in sensor networks. *IEEE Trans. Information Theory* 52 (7): 3349–58.

22. Zhang, Y., N. Ansari, and W. Su. 2011. Optimal decision fusion based automatic modulation classification by using wireless sensor networks in multipath fading channel. In *Proceedings of Global Telecommunications Conference (GlobeCom)*, 1–5. Piscataway, NJ: IEEE Press.

23. Xu, J. L., W. Su, and M. Zhou. 2011. Asynchronous and high-accuracy digital modulated signal detection by sensor networks. In *Proceedings of Military Communications Conference (MILCOM)*, 589–94. Piscataway, NJ: IEEE Press.

24. Xu, J. L., W. Su, and M. C. Zhou. 2011. "Likelihood Ratio Approaches to Automatic Modulation Classification," *IEEE Trans. on Systems, Man, and Cybernetics: Part C*, 41(4), pp. 455–469, July 2011.

Section III

Application Experiences

9 Challenges in Wireless Chemical Sensor Networks

Saverio De Vito

CONTENTS

9.1 INTRODUCTION

The capability to detect and quantify chemicals in the atmosphere, together with their concentrations, is a source of valuable information for many safety and security applications, ranging from pollution monitoring to detection of explosives and drug factories. Several gases are considered responsible for respiratory illness in people, and some of them (e.g., benzene) are known to induce cancers in the case of prolonged exposure, even at low concentrations [1]. Volatile organic compounds (e.g., formaldehyde) released as off-gases by furniture adhesives, by cleaning agents, or by smoking in indoor environments can reach concentration levels that are orders of magnitude higher than in outdoor settings. Hazardous gases, like explosives or flammable gases, are also sources of increasing concern for security reasons due to their possible use in terrorists attacks on military or civil installations. Moreover, some of them are currently in use or are foreseen to be used as energy carriers for automotive transport, and so their diffusion is expected to grow significantly. For example, hydrogen-powered car refilling stations could become very common in the near future [2]. The estimation of chemical distributions is hence significantly relevant for citizens' safety and security, and sensors capable of detecting such chemicals could prove to be life-saving assets. Chemical sensors are also recognized as a technological enabler for the emerging "smart cities" concept. In this concept, the development of human and social capital together with traditional and ICT infrastructures is at the basis of a new assessment of quality of life and sustainable productivity of a city. In this sense, distributed chemical sensors, both fixed or mobile (e.g. carried on board by bus or embedded in personal computing devices with a participating approach) will help to assess both pollutant distribution and personal exposure to the benefits of all citizens.

However, chemical monitoring in both outdoor and indoor environments is significantly affected by the peculiarity of the chemical propagation process [3]. To be concise, fluid dynamic effects like diffusion and turbulence, as the main drivers of propagation, make it very difficult to predict gas concentrations in the space and time domains. A single point of measure is usually totally ineffective, emphasizing the need for distributed approaches to detection of chemicals and estimation of their concentrations. In many applications, this requires the monitoring task to be fulfilled by a network of wireless (sometime mobile) modules, with the wireless term being related to either connectivity/communication and/or power supply. These architectures are called *wireless chemical sensor networks* (WCSN). For example, the plume generated by an H_2 spill in a hydrogen-based car refilling station could move along rather unpredictable paths, and the probability of a fixed, single, solid-state chemical sensor being hit by the plume with a significant concentration in a timely way could be negligible in many circumstances. At the same time, even if it could detect the plume, it would be impossible to assess the position of the spill. A distributed WCSN could dramatically improve the detection chance, while an autonomous moving sensor exploring the station would have the best chance to locate the spill.

When it comes to pollution monitoring, the commercial state of the art proposes the use of conventional spectrometer-based stations, which are large and expensive. The high costs would make it very difficult to achieve the required density in the measurement mesh of a cityscape to obtain statistically significant estimations in a complex environment (e.g., the influence of canyon effects). As a consequence, public policy makers are unable to make appropriate mobility management decisions to avoid or mitigate air pollution. The use of low-cost, compact, and wireless multisensory devices can offer a viable solution. Furthermore, because of their limited footprint, they can be deployed almost everywhere, including cities' historical centers, without altering the landscape of cultural heritage.

In this framework, researchers recently began to tackle the three-dimensional (3-D) unconstrained chemical-sensing scenario with two approaches. The first is based on the use of a moving detector. Together with appropriate modeling information, these detectors can follow random or goal-oriented paths, exploring a particular environment before being hit by a chemical plume [4, 5]. After that, by using chemical-spills search algorithms, often exploiting biomimetic approaches, they try to detect the source of contamination in the so-called source-declaration problem. The other approach basically relies on the use of multiple, low-cost, and autonomous distributed fixed detectors that try to cooperate in reconstructing a chemical image of the sensed environment [6, 7].

Advantages of the networked approach are identifiable in flexibility, scalability, enhanced signal-to-noise ratio, robustness, and self-healing. Several sensor nodes can be placed in different locations, each one with its own characteristics in terms of environmental conditions (air flow, temperature, humidity, different gas concentrations, etc.) that contribute to describe more thoroughly the environment in which they are embedded. Each smart chemical sensor, composing the distributed architecture, has its own communication capabilities, and its information is available to more than one client. The network can adapt itself to a variable number of chemical sensors, thereby improving reliability. If a sensor fails, the network can estimate its response on the basis of the previous behavior and of the response of the closest sensors while being able to self-heal the network structure by reconstructing routing trees [8]. On the other hand, an autonomous moving robot guarantees enhanced coverage and could be more effective in the source-declaration problem. Of course, the two approaches may be combined in the use of an autonomous fleet of chemical-sensing robots.

Nevertheless, engineers trying to design real-world operating WCSN systems face a number of challenges that limit their practical application. The most common challenges in both indoor and outdoor settings include designing appropriate strategies for single-module calibration and sensor stability, calibration transfer, efficient power usage, and 3-D mapping of gas concentration.

In this chapter, we will review these challenges together with the solutions proposed by our group during our work on the topic of wireless chemical sensing.

9.2 MODULE CALIBRATION AND SENSOR STABILITY

The characteristics of an optimal sensor for distributed chemical sensing should include low-power operations capability, low cost, long-term reliability, and stability; it should also be easy to integrate with simple signal-conditioning schemes. Depending on the application, the sensor should possibly express good specificity properties, high sensitivities, and very low detection limits. By far, this depiction applies more to an ideal device than a real one; in fact, no current chemical sensor technology seems near to obtaining such results simultaneously.

Most chemical sensors suffer from nonspecificity, i.e., their response to the target gas is heavily affected by so-called interferents. Indeed, interferents produce response changes that are indistinguishable from the one induced by a target gas, hampering detection and quantification capabilities. Chemical multisensor devices, also called electronic noses (e-noses), practically exploit the partial overlapping responses of an array of nonspecific sensors to detect and estimate concentrations of several gases simultaneously. Often this capability extends to gases toward which none of the enrolled sensors have been targeted. In this case, the response of a sensor subset can provide information by exploiting partial specificities or, more rarely, a peculiar ratio among chemical species in fixed-ratio mixture scenarios (proxy sensing).

Due to the intrinsic complexities of model-based approaches, calibration is mainly achieved by the use of multivariate techniques based on statistical regressors like artificial neural networks (ANNs). In this case, a data set containing the response of the sensor array to a representative subset of gas concentrations can be used to train a statistical regressor to detect and estimate gas concentrations in mixtures. Unfortunately, when it comes to complex mixtures and dynamic environments like the ones involved in monitoring city air pollution, it is nearly impossible to generate such a representative data set with synthetic mixtures and laboratory measurements, strongly limiting the application of solid-state multisensory devices in this framework.

In 2007, we began investigating the use of field-based calibration methodologies that exploit field-recorded samples captured from multisensor devices and colocated conventional stations to build adequate data sets for the training of ANN-based models [9]. In fact, we have investigated the use of a number of samples belonging to intervals ranging from 1 day to 96 days to build representative data sets. Results shows that a relatively compact data set recorded during 10 consecutive days allows the trained regressor to achieve rather interesting performance when estimating the concentration of several pollutants for a significant number of months to come. In practice, using such methodology for near-real-time estimation of benzene concentration could be performed with a mean relative error (MRE) of less than 4% (see Figure 9.1). However, it was evident that the combined effects of sensor drifts and

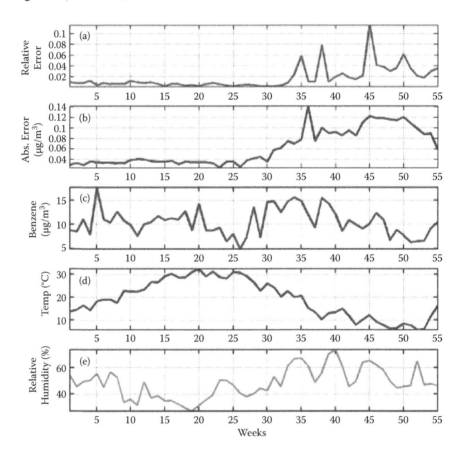

FIGURE 9.1 Benzene concentration estimation results in the 10-day training length run (weekly averages): (a) MRE; (b) MAE; (c) benzene concentrations as measured by a conventional station; (d) temperature measured by multisensor device; (e) relative humidity measured by the multisensor device. Results are affected by what we expect to be seasonal meteorological effects, evident after the 30th week (starting in November) superimposed to a slow degradation due to sensor aging effects (drifts). After the 50th week, the absolute error shows a definite recovery trend.

concept drifts induced by seasonal cycles and human activities caused the calibration to lose accuracy over time. Concept drifts are defined as a change in the target variables value probability distribution function that occurs in time as opposed to sensor drifts that relate to sensor response function changes during time to poisoning, aging or uncontrolled environmental dependencies. Sensor recalibration would be the obvious remedy, but because of the number of deployed multisensory devices is very high, conventional recalibration procedures employing reference gases or field-based approaches cannot be easily implemented. State-of-the-art drift-correction algorithms, which subtract the contribution of drift to sensor responses, can provide very interesting results, but these need a significant number of samples themselves in order to ground the value of their free parameters.

We recently began to investigate the use of semisupervised learning techniques as a way to adaptively solve this problem, which could be classified as a dynamic pattern-recognition problem (see Figure 9.2) [10]. In fact, distributed architectures are often built with sensing nodes that sport relevant but unused computing capabilities. This could provide diagnostic services and thus provide the possibility to self-improve the stability of sensor metrological characteristics. In particular, this could be significant for implementing drift-adaptation strategies, thereby accommodating this paramount problem in solid-state chemical sensing.

In networked configuration, sensor recalibration could also be performed cooperatively by temporarily adding other moving-reference sensors or, thanks to data-fusion techniques, using mutual recalibration strategies. Practically, the development of application-specific algorithms could allow mutual calibration by exploiting

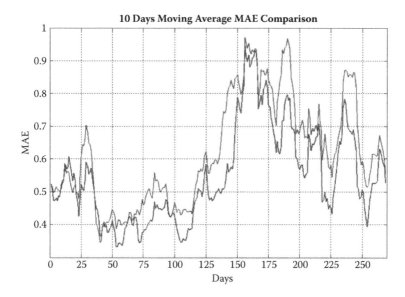

FIGURE 9.2 Drift counteraction with semisupervised learning. CO estimation comparison with secure socket layer (SSL) algorithm (red) and standard neural network (NN) algorithm (green) based on only 24 training samples (1 day). The SSL approach achieved an 11.5% performance gain with respect to the 1-year-long averaged MAE score.

networked cooperation in a totally unmanned fashion. In a work from Tsujita et al. [11], a network of NO_2 sensors recalibrated their baseline response by exploiting the identification of baseline conditions (very low NO_2 concentrations) through cooperative sensor responses under a particular set of conditions (dawn at low humidity and reference temperature) [11].

In any case, sensor drift remains a significant issue that prevents the spread of the use of multisensing devices in urban pollution scenarios. The future direction of research includes a combination of advanced signal-processing techniques and improved sensing capabilities, thereby reducing both nonspecificities and sensor drift. Calibration-transfer strategies are also to be developed to cope with sensor-production issues like inhomogeneity.

9.3   ENERGY EFFICIENCY IN WCSN

Any effort to develop a battery-operated WCSN must address the issue of power consumption. State-of-the-art metal oxide chemical sensors (MOX) require high working temperatures for the best sensitivity and specificity [12]. This issue limits the possibility to use them in wireless chemical sensing motes because their average power consumption is in the range of hundreds of milliwatts with continuous operation, allowing only a limited life span in battery-operated nodes. In contrast, polymer-based chemiresistors, resonators, and mass sensors (QMBs, i.e., quartz micro balance sensors or SAWs, i.e., surface acoustic waves sensors) are usually operated at room temperature [13]. Although they are not as common as their MOX counterparts, their low-power operation capability can be recognized as a huge advantage with respect to the other technologies, especially where detection of VOCs (volatile organic compounds) is concerned [14, 15]. Unfortunately, they may not be as efficient at very low concentrations, and most of them are significantly affected by humidity and are prone to heavy drift [16].

In the past decade, the reactivity of polymer/nanocomposites to chemicals has also been applied to the development of passive resonant sensors that are capable of wireless remote operations in a very simple way. LED/polymer-based optical sensing can be an extremely interesting solution for all applications where high limits of detection are not an issue because of their very low cost, reliability, and very low power demand [17]. However, to obtain suitable sensitivity in most applications, laser sources should be used, and this may significantly increase both costs and power needs.

From an architectural point of view, current commercial e-noses are not designed to tackle distributed chemical sensing problems, especially with regard to power management; however, a number of novel approaches have recently been proposed and experimentally tested. This field is hence rapidly evolving to exploit the plethora of results obtained by researchers in the field of wireless technology. For example, Bicelli et al. [18] investigated the use of commercial low-power MOX sensors in a WSN network for indoor gas-detection applications. They first suggested the use of a specially designed pulsed heating procedure in an attempt to achieve a significant increase in the wireless sensor battery life (about 1 year) with sample periods in the range of 2 minutes. Their results showed a serious trade-off between total power consumption and actual response time. Pan et al. [19] realized a single wireless electronic nose for online monitoring of livestock farm odors by integrating meteorological information and wind vector; however, detection performance and power consumption were not reported, so it was not possible to estimate autonomous life expectation. Becher et al. [20] recently presented a four-MOX sensor-based WCSN network for flame detection in military docks, but again, power needs restricted the application to the availability of power mains.

In recent years, our work on WCSNs has focused on the development of a wireless e-nose platform called TinyNose [21], based on commercially available WSN motes and controlled by software components relying on the TinyOS operating system [22]. The platform makes use of four low-power polymeric chemical sensors operating at room temperature that extend the entire sensory network life span at the cost of a low sensitivity. When using low-power sensors, radio power becomes significant in the overall power budget of the platform. In a 2011 work [23], we focused on the possibility of using computational intelligence algorithms to further extend the network life span by implementing a sensor-censoring strategy [24–26] focused on allowing the transmission of informative data packets while neglecting uninformative data. The target scenario for the application of the proposed methodology was the distributed monitoring of indoor air quality and, specifically, the detection of toxic/dangerous VOC spills. The same architecture has been used to develop and test an ad hoc algorithm for real-time 3-D gas mapping with a WCSN in simulated indoor environments in an effort to extract the needed semantic content from the deployed network [27].

Specifically, our platform relies on the use of nonconductive polymer/carbon-black sensors whose sensing mechanism of response is, at its simplest level, based on swelling. When the isolating polymer film, within which conductive particles have been dispersed, is exposed to a particular vapor, it swells while absorbing a varying amount of organic vapors, depending on the polymer's type. The swelling disrupts conductive filler pathways in the film by pushing particles apart, and the electric resistance of the composite increases [15].

To obtain a suitable voltage signal, i.e., one showing proportionality to variation in sensor resistance, a simple amplified resistance-to-voltage converter signal-conditioning system was implemented. Suitable choice of circuit parameters allows the proper operation of the board within a wide range of base resistance. Figure 9.3 shows a simplified functional scheme of the board.

The overall sensing subsystem was then directly connected to four analog-to-digital (A/D) input ports provided by the core mote of the proposed platform, the commercial Crossbow TelosB mote, a research-oriented mote platform that has shown its operative potential over time [28]. Powered by two AA batteries, the chosen platform is based on a low-power T I MSP430F1611 RISC microcontroller featuring 10 kB of RAM, 48 kB of flash, and 128 bytes of information storage. The embedded low-power radio, Chipcom© CC2420, represents the kernel for the IEEE 802.15.4 protocol's stack support.

The heterogeneity of the practical applications and the limited energy availability required a careful design of the software structure and protocols for the management of the wireless e-nose platform. The TinyOS operating system guaranteed the ability to keep the focus on domain-specific optimization and, in particular, on the implementation of the sensor-censoring strategy.

TinyOS is an open-source operating system for WSN applications developed by the University of California at Berkeley [22]. Essentially, its component library

includes network protocols, distributed services, basic sensor drivers, and data-acquisition tools, all of which can be further refined for custom applications. Specific software modules allow for the basic management of the sensor node, like local input/
output (I/O) and ratio transmissions. Furthermore, the power-management model
of TinyOS allows for the automatic management of module subsystems switching
between active, idle, and sleep phases to achieve a generalized power-management
strategy.

A C-derived language, NesC, is the reference programming language for TinyOS-
based programming. NesC relies on a component-based programming module.

Figure 9.3 shows a Unified Modeling Language (UML) deployment diagram
of the overall platform software architecture from which its three-layer design is
clearly apparent. The first layer encompasses all the embedded software components, developed in NesC that provide local control of the pneumatic section of the
electronic nose (if available), sensor interface control, data acquisition and processing, and eventually data transmission capabilities. At the second layer, a PC-based
component coded in Java captures data packets from a sensorless node that acts as a
gateway toward the IP-based network and storage facilities. At the third level, multiple GUIs (graphical user interfaces), also coded in Java, can provide visualization
and recording features while remotely controlling relevant parameters for embedded
software operations (e.g., duty-cycle parameters). Actually, the overall architecture
has been designed to host two pattern-recognition and sensor-fusion layers. The first
layer defines methods to connect a sensor's raw-data-processing component that will
allow for local situation awareness. Actually, it is the connection to the local sensor

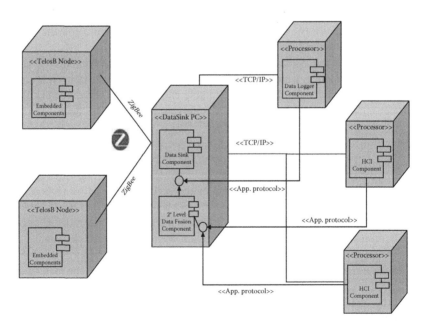

FIGURE 9.3 UML-deployment diagram of the software architecture for the proposed
platform.

fusion component that allows for the local estimation of pollutant concentration by using a trained neural network algorithm. The second layer provides second-level sensor-fusion services, thereby integrating estimates from all of the deployed e-noses. This level is responsible for the cooperative reconstruction of an olfactory image of the environment in which they are deployed.

9.4   DUTY CYCLE AND POWER CONSUMPTION

The TinyNose node has been designed for continuous real-time monitoring of volatiles with a programmable duty cycle, including the sensor's data acquisition, processing, and transmission toward a data sink. Duty-cycle parameters—and in particular the length of each phase and the sampling frequency—are fully programmable by the application designer, and sampling frequency can be set dynamically at run time through the GUI. To fully characterize the power consumption, we can easily separate the duty cycle in four separate phases, each one having different power needs:

Sleep phase: In this phase, each node is put asleep. MCU (micro controller unit) and radio are turned to standby mode.

Sensing phase: Sensors driving, data acquisition, and A/D conversion are carried out.

Computing phase: Data processing is performed to prepare data to be transmitted toward a data sink.

Transmission/reception phase: Actual data transfer occurs.

Battery life can be optimized by controlling the duration of each phase and the activation of the sensors' driving electronics.

At any instant, a single-module power consumption can be computed as a function of its microcontroller power state and whether the radio, pump, and sensor-driving electronics are on, and what operations the active MCU subunits are performing (analog to digital conversion). By using appropriate programming models, e.g., relying on split-phase operations, it is possible to best utilize the features provided by TinyOS to keep the power consumption to a minimum. In particular, as regards radio stack management, we chose to rely on the low-power listening (LPL) algorithm [22]. As such, the node radio can be programmed to switch on periodically just long enough to detect a carrier on the channel. If a carrier has been detected, then the radio remains on long enough to detect a packet to be routed to the data sink for mesh-shaped networking. After a time-out, the radio can be switched off.

To assess the base consumption of the TinyNose platform, a measuring setup based on a Tektronix TDS 3032 digital oscilloscope was set to measure Vshunt on 10-Ω shunt resistance and to let us derive current. Actually, in its simplest configuration, with sample frequency set at 1 Hz, the sensor node, as mentioned previously, remains in sleeping mode for most of the time (T_{RS} time interval). In this phase, a current I_{RS} measured as 12 µA, is drawn. During the T_{RW} wake-up period,

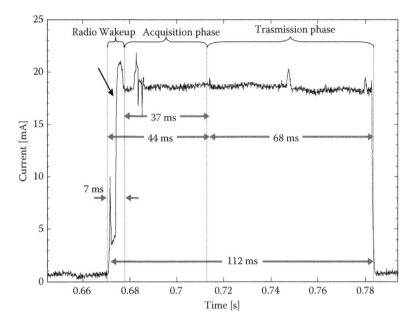

FIGURE 9.4 Current absorbed by TelosB in its active operating phase—in particular during the radio wakeup, sensing, and transmission phases in a complete cycle.

the sensor node turns on the chip radio, awakening from the sleep mode. In this stage we measured a 5-mA mean current draw. After that, the system set itself up for data capture and conversion with a mean current draw of 19 mA. Even in this stage, provided that data are sent every acquisition cycle, radio activity is the main source of power consumption. In fact, power consumption, in the active phase, is dominated by radio activity until the occurrence of a switch-off time-out inset; after an A/D converter time-out, it can be measured as 18 mA. Eventually the RF (radio frequency) circuit is turned off again, and the state changes from transmission mode to sleeping mode.

Figure 9.4 show the detailed evolution of the TelosB-drawn current during acquisition, processing, and transmission of sensor data, while Table 9.1 shows the

TABLE 9.1

Core Mote Consumption in Different Phases of the Duty Cycle

Operation phase	Time (ms)	Current request (mA)
Radio sleep (RS)	$T_{RS} = 888$	$I_{RS} = 0.012$
Radio wakeup (RW)	$T_{RW} = 7$	$I_{RW} = 5$
Acquisition (A)	$T_A = 37$	$I_A = 19$
Computing (C)	$T_C = 25$	$I_C = 2.5$
Transmission (T)	$T_T = 68$	$I_T = 18$

TABLE 9.2

Platform Signal-Conditioning-Board Consumption

Operation phase	Time (ms)	Current request (mA)
Operating condition	$T_{on} = 30$	38

consumption in terms of measured currents during each operating phase. Because of the high consumption of the prototype signal conditioning board (38 mA, as seen in Table 9.2), a digital signal drives a switch that lets the electronics board power supply to be switched on only during the data-capture time slice. This strategy is made possible by a polymer-sensing mechanism; in fact, polymer swelling is only negligibly affected by actual sensor power-up, so that the sensors do not need warming up before their resistance could be sampled.

Usually, module life span can be estimated using the mean current consumption, namely $I_{cc,mean}$, of the wireless sensor considering battery capacity, conversion efficiency, and power-supply output voltage gain.

The mean current value, $I_{cc,mean}$, obtained under the basic operating conditions without applying sensor censoring, can be computed according to Equation (9.1) by using the current measured in the proposed setup:

$$I_{cc,mean} = \frac{T_S}{T} I_{RS} + \frac{T_{RW}}{T} I_{RW} + \frac{T_A}{T} I_A + \frac{T_T}{T} I_T. \tag{9.1}$$

Neglecting conversion efficiency, battery life (BL) can be computed as a function of the battery capacity C and total current draw. For our w-nose, C is equal to 3500 mAh, while total current is equal to the sum of mean current absorbed by the conditioning board $I_{b,mean}$ (30/1000 × 38 = 1.14 mA) and the $I_{cc,mean}$ (1.97 mA computed using Equation [9.1]). Expressing BL as the ratio $C/(I_{cc,mean} + I_{b,mean})$, we can estimate that the overall e-nose battery life, with sampling and transmission frequency of 1 Hz, is roughly 47 days, a rather interesting value for a four-sensor four-nose. Extending the sample period to 10 s, hence losing real-time characteristics, this basic setting will account for a battery life in excess of 1 year.

9.5   POWER SAVING USING SENSOR CENSORING

The additional power required for the execution of the local data-processing components should be seriously taken into account in order to evaluate the benefits of censoring strategies. However, the outcome depends basically on the rate at which significant events occur. In most chemical-sensing scenarios, the probability p of a significant event to occur (e.g., chemical spills) is expected to be very low, while the timely transmission of relevant data is needed in security applications.

In order to perform an experimental check of the sensor censoring concepts in WCSN scenarios, we designed and implemented an ad hoc lab-scale experiment. A sensor-fusion component that was trained for a distributed pollutant-detection

TABLE 9.3

Memory Footprint Increase in the Embedded Component Resulting from the Linking of Sensor-Fusion Neural Component

Algorithm	Bytes in ROM	Bytes in RAM
Basic	20,380	574
Basic+NN comp.	27,340	910

application was developed, and the power savings obtained by using sensor censoring were evaluated. Actually, we assumed a general chemical-sensing problem characterized by the presence of two pollutants whose toxic/dangerous concentration limits were different. To save battery energy, the single mote should be able to decide whether or not to transmit the sampled data on the basis of the concentration of the two gases estimated by sensor responses. The e-nose sensor arrays were exposed to different concentrations of acetic acid and in a controlled environment setup, and their responses, sampled by the motes, were recorded to build a suitable data set for the training of an ANN component to be run onboard. ANN architecture was chosen based on considerations about its flexibility, high capacity, compact knowledge representation (low-space footprint), and low computational demands (see Tables 9.3 and 9.4). The onboard component was then loaded with network weights obtained by training an identical model in MATLAB with the recorded data set so as to reach a reasonable point-to-point real-time estimation of concentrations of different analytes.

The proposed architecture performance was evaluated by dividing the validation-set mean absolute error (MAE) by the analyte concentration's range span, obtaining a 6% (standard deviation = 10%) value for the acetic acid concentration-estimation problem and 11% (standard deviation = 11%) for the ethanol problem (see Figure 9.5).

Following a measurement approach, the execution of the ANN (artificial neural network) sensor-fusion component accounted for an additional 2.5-mA consumption over a total time span of 25 ms (function call overhead included).

TABLE 9.4

Computational Complexity in Terms of Functional Calls

Functional	Calls per Each NN Estimation
Tanh	10
Multiplication	76
Sum	82

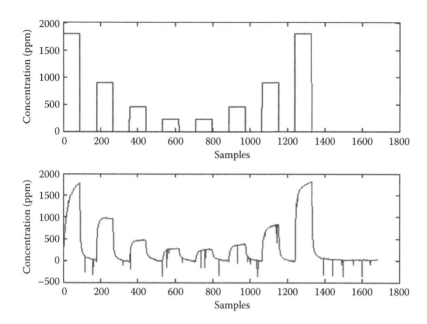

FIGURE 9.5A   Acetic acid concentration estimation (red) performed by the feed-forward NN (FFNN) component plotted against true concentration (blue). The x-axis depicts time (samples), while the y-axis depicts real and estimated concentrations values.

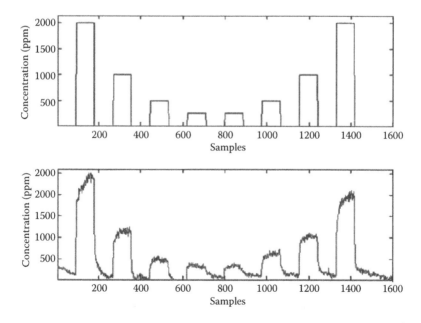

FIGURE 9.5B   Ethanol concentration estimation (red) performed by the FFNN component plotted against true concentration (blue). The x-axis depicts time (samples), while the y-axis depicts real and estimated concentrations values.

To reassess total power consumption, we rewrite Equation (9.1) by taking into account the computing phase.

$$I_{cc,mean}^{NN} = I_{RS}\left(1 - \frac{T_{RW} + T_A + T_T + T_C}{T}\right) + I_{RW}\frac{T_{RW}}{T} + I_A\frac{T_A}{T} + I_T\frac{T_T}{T} + I_C\frac{T_C}{T}. \quad (9.2)$$

This equation provides the mean current supplied by the batteries in the case when a significant event occurs: The mote performs signal conditioning and data sampling, data fusion by means of NN and data transmission every T seconds.

Using a Bernoulli random variable $X \approx B(1, p)$, which would model the result of the computation performed by the NN to verify whether to transmit the acquired data on the radio channel, Equation (9.2) can be rewritten as

$$I_{cc,mean}^{NN} = I_{RS}\left(\frac{T'_{RS}p + T''_{RS}(1-p)}{T}\right) + I_{RW}\frac{T_{RW}}{T}p$$

$$+ I''_A\frac{T_A}{T}(1-p) + I'_A\frac{T_A}{T}p + I_T\frac{T_T}{T}p + I_C\frac{T_C}{T}, \quad (9.3)$$

where T'_{RS} and T''_{RS} are, respectively

$$\begin{cases} T'_{RS} = T - T_{RW} - T_A - T_T - T_C \\ T''_{RS} = T - T_A - T_C \end{cases}. \quad (9.4)$$

By exploiting Equation (9.3), we can finally discuss the advantages of NN-based sensor censoring for the implementation of power-saving strategies in the proposed architecture.

The worst case is obtained when $p = 1$, i.e., when all samples refer to significant events; in that case, the mean current computed using Equation (9.3) is obviously greater than the one calculated by Equation (9.1), i.e., $I_{cc,mean}^{NN} = 2.013$ mA. However, by equating Equations (9.1) and (9.3) and selecting p as an independent variable, we obtain

$$p = 0.97. \quad (9.5)$$

The obtained value represents the percentage threshold of a significant event under which the NN-based sensor censoring becomes more efficient in power management. This computed threshold level makes the proposed approach feasible for most of the analyzed distributed chemical-sensing scenarios.

In monitoring industrial chemical spills, even considering false-positive generation, it is reasonable to expect values of p that reflect a number of only a few significant samples per day. Of course, the censoring criteria, i.e., the "spiking" threshold, should be chosen after exploring the trade-off between sensitivity and the false-positive rate caused by estimation errors while also considering the danger/toxicity level of the target gas. For an experimental evaluation, using the previously mentioned setup and allowing a small slack to the network estimation of ethanol (detection threshold on NN response = 10 ppm with respect to a 2000-ppm max exposure level), we obtain a false positive rate of 5% for ethanol and less than 1% for acetic acid. In the case of no positive events during a node's lifetime—a reasonable expectation in a safety-oriented leakage-detection scenario—a proposed node can be expected to provide a maximum operative life span that is very near to the intrinsic limit now dominated by the power needed by the signal-conditioning board. In particular, considering $p = 0.01$, the expected lifetime computed by taking Equation (2.3) into account reaches 113 days (110 days for $p = 0.05$), which is a rather interesting value for a four-sensor wireless electronic nose with real-time operating characteristics.

9.6   3-D RECONSTRUCTION

As stated in the introduction (Section 9.1), the capability to reconstruct 3-D gas mappings is becoming more and more interesting for possible applications in city air-pollution mapping, localization of hazardous gas spills, and energy efficiency (e.g., efficient control of HVAC automation in buildings). Recently, we investigated the possibility of obtaining 3-D quantitative indoor-air quality assessment with the TinyNose architecture [27]. The outcome revealed the capability of building gas-concentration mapping of acetic acid–ethanol mixtures (as VOC pollutant simulants) in ambient air with the mesh of four w-noses coupled with a two-stage sensor-fusion system. To focus on the 3-D reconstruction problem, a set of four TinyNose w-noses equipped with a four-MOX sensor array was considered. A basic performance esti-mation procedure was conducted on a single TinyNose equipped with the novel MOX-based sensor array. Using a controlled climatic chamber, the node was exposed to different concentrations of acetic acid and ethanol at different relative humidity percentages (see Table 9.5). By using sensor responses, a two-slot tapped-delay neu-ral network (TDNN) with 10 hidden-layer neurons for instantaneous estimation of concentrations obtained an MAE/range value of 2.34% (0.75 ppm) for acetic acid and 6.5% (9.8 ppm) for ethanol (see Figure 9.6).

However, following a field-based approach to calibration, we decided to design a second experimental setup involving the deployment of a network of four instances of TinyNose in an ad hoc glass box (volume = 0.36 m³) and evaluate their capability to reconstruct a real-time 3-D chemical concentration image of the two previous pollutants. Different amounts of the two chemicals (see Table 9.6) were introduced

TABLE 9.5
Test Conditions for Experiments in Controlled Chamber

	Gas Concentration Ranges	
RH (%)	Acetic Acid (ppm)	Ethanol (ppm)
[20, 30, 50]	[0,5,7,10,15,20,25,30,32]	[0,15,30,70,90,115,130,150]

Note: The different gas mixtures used in this experiment were synthesized using all com-binations of acetic acid and ethanol at the given relative humidities for a total of 216 cycles. The baseline mixture was set at RH = 50%.

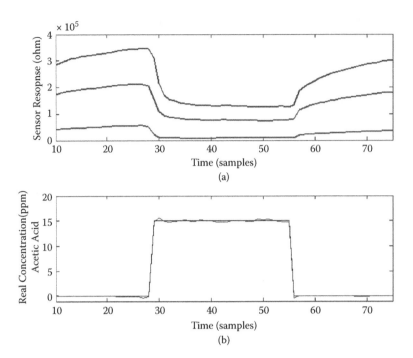

FIGURE 9.6 (a) Sensor-array responses in controlled chamber setup; (b) trained TDNN responses and ground truth comparison during the corresponding complete-exposure cycle.

and diluted until complete evaporation with the use of a standard PC fan. Based on the reasonable hypothesis of uniform concentration distribution over the box, at steady state, the response of the node was sampled (1 Hz) to build a suitable data set for the training of the TinyNoses. This was based on a two-level classifier/regressor scheme. In this way, each of the w-noses was made capable of estimating

---

TABLE 9.6

Test Conditions for Experiments in Glass Box

Gas Concentration Ranges

Acetic Acid (ppm)	Ethanol (ppm)
[0,5,10,15,20]	[0,6,12,17,23]

Note: The different gas mixtures used in this experiment were synthesized using all combinations of acetic acid and ethanol. Two exposure cycles were executed for each combination. Steady-state sensor array response to each exposure cycle was recorded to build the training data set. Ambient RH values and temperatures (not controlled) were also recorded to be part of the onboard NN feature set.

---

the local pollutant concentration at its deployment location. Another 10 runs of steady-state sample acquisition with the same procedure were then used to build a suitable test set. The test set MAE, averaged for all four w-noses, reached 3.15 ppm and 4.36 ppm for ethanol and acetic acid, respectively. These figures allow us to locate the expected absolute error on the real-time local concentration estimation under a 5-ppm threshold. This value is valid for concentration estimation in a mixture and, thus, in the presence of interferents, provided that concentration levels have a slow variation rate.

In order to reconstruct a 3-D chemical image of the glass box, the kernel-DV algorithm (see Nakamoto and Ishida [4]), originally developed by the Lilienthal group for use with mobile robot acquisitions, was adapted for real-time cooperative sensor fusion. The algorithm basically uses a 3-D Gaussian kernel to propagate localized measurement to a 3-D environment based on confidence values, depending on the distance from the actual measurement points. Estimations based on the kernel propagation are balanced with a default averaged value (homogeneous gas distribution) by the use of a confidence value that is normalized by a scale factor. Eventually, should the confidence value fall to 0 (points located far from all actual measurement points), the algorithm reverts to the homogeneous gas distribution hypothesis. In the last setup, 17 μg of ethanol was allowed to evaporate within the glass box in one of the left-down box corners. By using the neural calibration obtained previously and encoded in the onboard computational intelligence component, the single nodes were able to estimate local concentrations of both gases. Their estimations were transmitted and collected at a data sink, where a sensor-fusion component was coded to reconstruct an instantaneous 3-D chemical image of the box.

The overall performance of the 3-D reconstruction algorithm depends on the value of three base parameters of the kernel-DV algorithm—cell mesh width, the kernel width σ, and the confidence scale parameter—and, of course, the w-nose deployment positions. Cell mesh width only trades off 3-D reconstruction resolution with computational costs; for this reason, a fixed value that allowed for real-time reconstruction was selected. The confidence scale parameter depends on kernel width, so, for a fixed-deployment configuration of the w-noses, an automated procedure was designed to choose the appropriate kernel width parameter value on the basis of a leave-one-mote-out approach. Actually, in scanning a parameter-values array ([0.05, 0.10, 0.15, 0.2]), all but one sensing node were used to estimate the concentration values of the two analytes for each parameter setting, with the adapted kernel-DV algorithm used at the remaining node position. Figure 9.7a shows the comparison between local instantaneous concentration estimation at node position 3 and estimation carried out by the 3-D reconstruction algorithm at the same position. Without affecting generalization, the mean absolute difference between the estimated value and the actual value as estimated by the remaining node, during all the exposure time, has been defined as the performance value to be optimized by the brute-force approach.

Figure 9.7b shows the averaged instantaneous absolute difference among local and 3-D reconstruction-based estimations for the four motes in the ethanol case. Peaks can be spotted in the rise and fall of the concentration levels during transients.

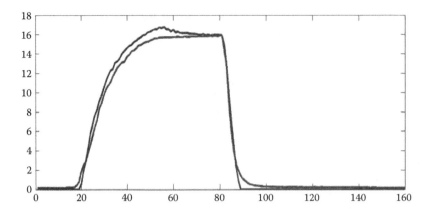

FIGURE 9.7A Instantaneous ethanol concentration estimation as computed at node 3 (blue) compared with the estimation obtained by 3-D reconstruction, using the remaining three nodes, at the same location. The experimental setup foresees the deployment of four w-noses in the 0.36-m^3 glass box and the release of 17 mg (20.9 ppm) of ethanol near one corner of the box.

The peaks can be explained by the different concentrations measured by the motes during transients due to their different positions. In this case, the peak magnitude could be effectively reduced by tuning mote positioning and the density of the measurement mesh in the sensed environment. Figures 9.8 and 9.9 depict, respectively, w-nose positioning and an instantaneous reconstruction of the concentration of the two pollutants. This preliminary result suggests the promise of using a w-nose deployment for real-time 3-D quantitative analysis of air quality in the presence of a pollutant mixture. Cross validation of motes has also been shown to be useful for tuning sensor-fusion algorithm parameters and in evaluating performance.

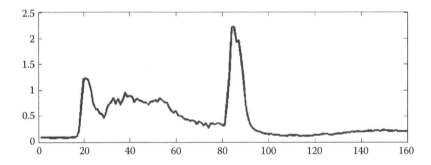

FIGURE 9.7B Instantaneous absolute difference among local ethanol concentration and 3-D reconstruction algorithm with kernel width σ set at 0.1. The instantaneous difference reported here was averaged throughout the leave-one-mote-out procedure executed for the exposure to 17 mg (20.9 ppm) of ethanol. The computed MAE value was used to optimize the 3-D reconstruction algorithm.

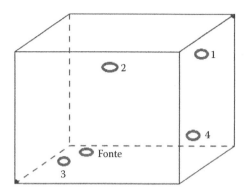

FIGURE 9.8   Positioning of the four w-noses and gas source within the glass box.

9.7   CONCLUSIONS

WCSN architectures can provide a new insight on important and potentially life-saving phenomena like the detection of pollutants and/or hazardous gas. Their practical diffusion is still limited by several issues like power management, sensor nonspecificity, and instability. Ad hoc 3-D diffusion models can be devised to obtain high-level semantic information about gas concentrations, with applications to localization of spills or to improving energy efficiency (HVAC automation).

In both cases, computational intelligence together with distributed sensing techniques may allow for viable solutions even in the case of using low-cost equipment exploiting redundancy and the power of machine learning. Since 2006, our research group, together with several researchers all around the globe, has been committed to the development of solutions for real-world application that could be engineered to the market. Despite the growing number of contributions in this field, the main

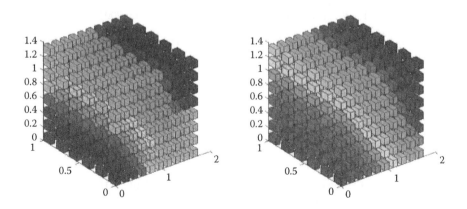

FIGURE 9.9   Instantaneous 3-D ethanol (right) and acetic acid (left) concentration reconstruction (computed at data sink) using a four w-nose deployment in the glass box experimental setup.

issues are still out there and appear to be solved only for specific segments of a complete application. Our efforts have produced a complete platform, called TinyNose, that addresses all of the applicable segments of WCSN architecture. The development of WCSNs is rapidly progressing toward real-world applications. However, massive field-test deployments are still needed to test their reliability in both indoor and outdoor scenarios.

ACKNOWLEDGMENTS

The core work for this chapter has been conducted by researchers at UTTP-MDB lab of the ENEA C.R. Portici. For cooperation and support, I am indebted to my coauthors and in particular to E. Massera, G. Fattoruso, G. Di Francia, and M. Miglietta.

REFERENCES

1. Dockery, D., C. A. Pope, X. Xu, F. Speizer, and J. Schwartz. 1993. An association between air pollution and mortality in six U.S. cities. *N. Engl. J. Med.* 329:1753–59.
2. Grimes, C., K. G. On, O. K. Varghese, X. Yang, G. Mor, M. Paulose, E. C. Dickey, and C. Ruan. 2003. A sentinel sensor network for hydrogen sensing. *Sensors* 3:69–82.
3. Nakamoto, T., and H. Ishida. 2008. Chemical sensing in spatial/temporal domains. *Chemical Review* 108 (2): 680–704.
4. Lilienthal, A. J., A. Loutfi, and T. Duckett. 2006. Airborne chemical sensing with mobile robots. *Sensors* 6:1616–78.
5. Lilienthal, A. J., and T. Duckett. 2004. Building gas concentration gridmaps with a mobile robot. *Robotics and Autonomous Systems* 48:3–16.
6. De Vito, S., E. Massera, G. Burrasca, A. Di Girolamo, M. L. Miglietta, G. Di Francia, and D. Della Sala. 2008. TinyNose: Developing a wireless e-nose platform for distributed air quality monitoring applications. In *Sensors, 2008 IEEE*, 701–4. Piscataway, NJ: IEEE Press.
7. Shepherd, R., S. Beirne, K. T. Lau, B. Corcoran, and D. Diamond. 2007. Monitoring chemical plumes in an environmental sensing chamber with a wireless chemical sensor network. *Sensors and Actuators B: Chemical* 121 (1): 142–49.
8. Akyildiz, I. F., W. Su, Y. Sankarasubramaniam, and E. Cayirci. 2002. Wireless sensor networks: A survey. *Computer Networks* 38:393–422.
9. De Vito, S., E. Massera, M. Piga, L. Martinotto, and G. Di Francia. 2008. On field calibration of an electronic nose for benzene estimation in an urban pollution monitoring scenario. *Sensors and Actuators B: Chemical* 129 (2): 750–57.
10. De Vito, S., G. Fattoruso, M. Pardo, F. Tortorella, and G. Di Francia. 2012. Semi-supervised learning techniques in artificial olfaction: A novel approach to classification problems and drift counteraction. *Sensors Journal, IEEE* 12:3215–24.
11. Tsujita, W., A. Yoshino, H. Ishida, and T. Moriizumi. 2005. Gas sensor network for air-pollution monitoring. *Sensors and Actuators B: Chemical* 110:304–11.
12. Barsan, N., and U. Weimar. 2001. Conduction model of metal oxide gas sensors. *J. Electroceram.* 7 (3): 143–67.
13. Arshak, K., E. Moore, G. M. Lyons, J. Harris, and S. Clifford. 2004. A review of gas sensors employed in electronic nose applications. *Sensor Review* 24 (2): 181–98.
14. Quercia, L., F. Loffredo, B. Alfano, V. La Ferrara, and G. Di Francia. 2004. Fabrication and characterization of carbon nanoparticles for polymer based vapor sensors. *Sensors and Actuators B: Chemical* 100:22–28.

15. Severin, E. J., B. J. Doleman, and N. S. Lewis. 2000. An investigation of the concentration dependence and response to analyte mixtures of carbon black/insulating organic polymer composite vapor detectors. *Anal. Chem.* 72:658–68.

16. Ha, S. C., Y. Yang, Y. S. Kim, S. H. Kim, Y. J. Kim, and S. M. Cho. 2005. Environmental temperature independent gas sensor array based on polymer composite. *Sensors and Actuators B: Chemical* 108:258–64.

17. Diamond, D., S. Coyle, S. Scampagnani, and J. Hayes. 2008. Wireless sensor networks and chemo-/biosensing. *Chem. Rev.* 108 (2): 652–79.

18. Bicelli, S., A. Depari, G. Faglia, A. Flammini, A. Fort, M. Mugnaini, A. Ponzoni, V. Vignoli, and S. Rocchi. 2009. Model and experimental characterization of the dynamic behavior of low-power carbon monoxide MOX sensors operated with pulsed temperature profiles. *Instrum. Meas., IEEE Trans.* 58 (5): 1324–32.

19. Pan, L., R. Liu, S. Peng, Y. Chai, and S. X. Yang. 2007. A wireless electronic nose network for odours around livestock farms. In *Proc. 14th M2VIP*, 211–16. Piscataway, NJ: IEEE Press.

20. Becher, C., P. Kaul, J. Mitrovics, and J. Warmer. 2010. The detection of evaporating hazardous material released from moving sources using a gas sensor network. *Sensors and Actuators B: Chemical* 146 (2): 513–20.

21. De Vito, S., G. Burrasca, E. Massera, M. Miglietta, and G. Di Francia. 2009. Power savvy wireless e-nose network using in-network intelligence. In *Olfaction and electronic nose: Proceedings of 13th ISOEN, AIP Conference Proceedings*, vol. 1137, 211–14. Melville, NY: AIP.

22. TinyOS. n.d. http://www.tinyos.net.

23. De Vito, S., P. Di Palma, C. Ambrosino, E. Massera, G. Burrasca, M. L. Miglietta, and G. Di Francia. 2011. Wireless sensor networks for distributed chemical sensing: Addressing power consumption limits with on-board intelligence. *Sensors Journal, IEEE* 11 (4): 947–55.

24. Rago, C., P. Willett, and Y. Bar-Shalom. 1996. Censoring sensors: A low-communication rate scheme for distributed detection. *Aerospace and Electronic Systems, IEEE Trans.* 32 (2): 554–68.

25. Appadwedula, S., V. V. Veeravalli, and D. L. Jones. 2008. Decentralized detection with censoring sensors. *Signal Processing, IEEE Trans.* 56 (4): 1362–73.

26. Jain, A., and E. Y. Chang. 2004. Adaptive sampling for sensor networks. In *DMSN '04: Proc. First Workshop on Data Management for Sensor Networks*, Vol. 72 of *ACM International Conference Proceedings Series*, 10–16. New York: ACM.

27. De Vito, S., G. Fattoruso, R. Liguoro, A. Oliviero, E. Massera, C. Sansone, V. Casola, and G. Di Francia. 2011. Cooperative 3D air quality assessment with wireless chemical sensing networks. *Procedia Engineering* 25:84–87.

28. Polastre, J., R. Szewczyk, and D. E. Culler. 2005. Telos: Enabling ultra-low power wireless research. In *IPSN '05: Proc. 4th international symposium on information processing in sensor networks*, 364–69. Piscataway, NJ: IEEE Press.

10 Low-Power, Extensive Sensor Networks from the Wired Perspective

Alan R. Wilson

CONTENTS

10.1 INTRODUCTION

Much effort around the world is going into the development of monitoring systems for large expensive assets with a view to reducing the cost of maintenance, increasing their availability, and reducing or eliminating catastrophic failures. The periodic monitoring of mechanical systems, such as engines, valves and motors, has been common practice for many years, virtually since these systems came into use. Continual data logging of mechanical systems had to wait for the development of the appropriate technology, starting with mechanically based recorders moving to present-day electronic systems. These have been in use for some time to detect and predict mechanical faults. The equivalent continual monitoring for structural condition is a developing area and presents challenges in relating the collected data to the state of the structure.

Structural monitoring is envisaged to replace maintenance regimes, such as:

- Scheduled replacement, which often means the routine replacement of components with many hours of remaining life.
- Scheduled inspections, which are costly and time consuming. A number of nondestructive inspection techniques are often used to detect structural damage such as fatigue cracks and corrosion. Scheduled inspections may also involve strip-down and then reassembly to enable inspection of difficult-to-access locations. A particularly attractive aspect of permanently installed systems is the monitoring of inaccessible areas that currently require deconstruction of the structure to access them.
- Repair on failure, which can be very costly when the failure occurs in a remote location or, worse still, may lead to a catastrophic event.

The monitoring of mobile assets such as ships and aircraft poses extra challenges compared to the monitoring of static structures such as bridges and buildings. For aircraft, there can be severe weight and power restrictions with, in some cases, the need for extended operations with no onboard power source. Aircraft sensors and systems must also be highly reliable, since undetected faults can have catastrophic

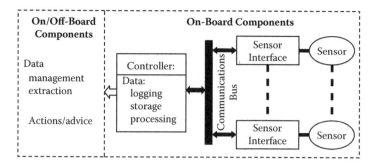

FIGURE 10.1 Components of a monitoring system split into parts that must be located on the structure and those that could be located off the structure.

consequences. Much of the work on structural monitoring systems has been done for applications on aircraft [1, 2], since these currently have very expensive inspection regimes required for safety considerations. For ships, the environment can be very challenging, and the number of sensors required can become quite large. This is particularly the case for complex structures that are used in a wide range of environments and operations, such as occurs for military ships. For example, a large civilian tanker is essentially a box structure that is operated in a routine manner. A relatively small numbers of sensors can be used to good effect to monitor the strains and deduce the loads on these vessels [3]. For comparison, although much smaller, an aluminum-hulled patrol boat with a semi-planing hull form has a much more complex structure and is operated over a wide range of speeds and operating environments. Thus the load distribution is complex and varies with the way that the boat is operated [4], requiring a greater number of sensors and more complex data processing to determine the loads on the structure.

The use of networks of distributed, embedded sensors provides a method to monitor large assets. A complete monitoring system consists of the following components, shown schematically in Figure 10.1:

- *Sensors* to transduce a physical/chemical event to a measurable signal. This is often an electrical signal, since these are easy to manipulate and record.
- *Sensor interfaces and signal conditioning* to interface to and/or amplify the signal from the sensor. Many modern sensors include signal conditioning and analog-to-digital conversion on the sensor and provide a digital output, so the interface in these cases must be in the form of a digital input (often serial data) and may also require a command-response protocol to acquire the data.
- *Data transmission and communication protocols* to transmit data to a central data-logging or data-collection point and to provide a means of remotely controlling the sensors on the network.
- *Data logging and storage.* The data storage requirements may become quite large for a system that is permanently on and include high-data-rate sensors such as strain gauges and accelerometers. For instance, over 1 month, 20

strain rosettes (which consist of three linear strain gauges at relative angles of 0, 45 and 90 degrees.) operating at 100 Hz make 16×10^9 measurements.

- *Data management* to manage large data sets as they are accumulated over time.
- *Information extraction*, since the collected measurements must be related to the actual structure. This could include the correlation of theoretical responses, derived from modeling of the structure, with the measured response of the structure.
- *Actions*, which would arise from the modeling and correlation of measurements to guide maintenance actions and, eventually, to predict required actions and manage structural capability.

This chapter will only consider systems that are permanently installed and require no human intervention in the measurement and recording of signals, i.e., embedded sensors, networks and data logging. In the context of structure monitoring, this is often referred to as *structural health monitoring* or as *embedded nondestructive evaluation* (NDE). Furthermore, this chapter will concentrate on the first four components of the previous list, since these must be embedded in the asset being monitored, whereas the other components can be managed externally to the asset. The systems developed at the Defence Science and Technology Organisation (DSTO) will be used as an example of aspects of a distributed sensor network. The other components of the system will be briefly touched on.

10.2   MONITORING SYSTEM NETWORKS

Networks of sensors can be organized in a number of ways. This section briefly considers some of the advantages and disadvantages of various ways of arranging sensors to monitor a structure.

10.2.1   STAND ALONE

In a stand-alone system, each sensor is an individual system with its own independent power and data storage.

Advantages:
- Easy to implement, since all sensors are independent; in particular, no wiring is required to interconnect sensors

Disadvantages:
- Manual retrieval of data
- Requires replacement of power sources (batteries) unless low-power and scavenging techniques can be used
- Not suitable for inaccessible locations
- Difficult to correlate time data from different sensors

The stand-alone system of sensors is only really suited to one-off trials and is not suitable for permanent installation.

10.2.2   WIRED: STAR NETWORK

In a wired star network, the sensors are all individually wired back to a central controller that performs all of the signal conditioning and measurement as well as the data recording.

Advantages:
- Simple to set up, simple protocols
- Off-the-shelf solution in many instances
- Power sourced from the controller
- Robust to individual sensor or wiring failure
- Flexibility, since either analog or digital communications can be used

Disadvantages:
- Individually cabled sensors, which can lead to issues regarding weight of cabling, long cable runs, and the thickness of cable bundles
- Unsuitable for extensive networks
- Difficult to retrofit
- Signal conditioning may be remote from the sensor, leading to noise and interference issues

Until recently, star networks were the most common form of network for monitoring mechanical systems and some structural monitoring. The automotive industry and industrial process monitoring have both moved away from star networks to multipoint networks, primarily to reduce cabling requirements. Likewise, the requirement to run cables to each and every structural monitoring sensor from a central controller is seen as a very limiting factor for the implementation of extensive sensor networks on a large vessel due to the cost, weight, and volume penalties of the cabling. Star networks of sensors are considered to be most useful for limited time trials with a limited number of sensors, where the controller is located close to the sensors.

10.2.3   WIRED: MULTIPOINT NETWORK

In a wired multipoint network, the sensors have local signal conditioning and computing/communications capability. A network bus provides power and communications to all of the sensors and is generally managed by a central controller.

Advantages:
- Much less cabling than the individually wired sensors
- Local signal processing leading to lower noise
- Digital network communications for data integrity
- Local computing capability for local data processing and storage and the potential for "smart" sensors that include local data processing, manipulation, and assessment

Disadvantages:
- More complex communications and data logging; requires some "smarts" at the sensor

- Difficult for retrofitting; the networking capability would be best installed during construction
- Susceptible to single-point failure compromising the whole network; Can be partly overcome by good design

Multipoint wired networks would appear to offer the best solution so far for implementing large arrays of sensors.

10.2.4   WIRELESS: MULTIPOINT NETWORK

A wireless network is similar to the multipoint wired network except that there is no cabling, and all communications are performed via wireless links. Various standards for radio frequency (RF) wireless communications protocols exist to achieve this.

Advantages:
- No wiring: less weight, easy to install
- Local signal and data processing
- Digital communications for data integrity
- Robust re single sensor failure
- Easiest to retrofit in some circumstances (open areas)

Disadvantages:
- Most complex communications requirements
- Requires manual replacement of power source (batteries) unless a millipower device is used that can scavenge energy from the structure (This can restrict the types of sensors used, e.g., metal foil strain gauges are high-power sensors.)
- Battery-powered system not suitable for inaccessible locations (This may also limit use over extended time scales.)
- Not suitable for multiple, conducting wall compartments due to RF screening; requires an open area to operate
- The use of RF may restrict use in some cases due to concerns re RF leakage (e.g., military platforms)
- Can suffer from RF interference but this can overcome by good design

10.2.5   DISTRIBUTED NETWORKS

A distributed network does not have a central controller or data store, these functions being dynamically distributed over the whole network, often with multiple communications paths. As such, distributed networks can only be implemented on multipoint wired or wireless networks. Data are stored throughout the network, so large local data storage capability may be required, and this is now possible with the advent of high-capacity flash memory. Nodes in a distributed network are often aware of their nearest neighbors but not of the whole network. The communication in a distributed network is sometimes performed by "software agents," a piece of software that acts on behalf of the program on one node, which are transmitted between nodes to communicate with programs on other nodes.

Advantages:
* Can be very robust re single point failures

Disadvantages:
* Very complex communications protocols
* Low-power operation becomes more complex to manage
* Time synchronization becomes more complex to manage, since there is no single controller
* Data extraction may be more time consuming, since large data sets may have to be funneled through the network to the data collection point rather than just swapping over a single mass-storage device

10.2.6 Conclusions for Monitoring System Networks

The wireless network solution has similar attractions to the multipoint wired network with the added benefit of no wiring. However, the requirements for manual battery replacement, or restrictions due to the limited power available by energy scavenging, and difficulties in RF transmission through conducting barriers (e.g., bulkheads in ships) count against wireless networking. Distributed networks are also currently seen to require too much overhead and complexity in managing the network, particularly with regard to achieving low-power operation and synchronization of the data collected over the network. The overall conclusion is that, at the present time, wired multipoint networks are the most practical method to implement extensive networks of sensors, particularly if the required cabling is installed during the construction of the vessel.

In practice, a sensor network could incorporate a mix of the previously discussed configurations. For instance, a network could consist of a number of multipoint wired networks (i.e., a star network of multipoint networks) under the control of a central controller. This would improve the robustness of the system and makes sense if the controller is centrally located so that a few networks radiating out from it will not introduce much extra cabling. Also, rechargeable power sources (e.g., batteries) may be required at sensor nodes in cases where the power to the sensors may be interrupted for extended periods of time. An example is the monitoring of corrosion on military aircraft. Unlike civilian aircraft, military aircraft can spend long times on the ground and with no power. Most corrosion in these aircraft occurs on the ground, where there is moisture and condensation. Thus any corrosion sensors would need to operate autonomously during these periods. Fortunately, corrosion measurements can be made infrequently and can use low-power sensors so that battery power would be sufficient for extended periods of operation.

10.3 MONITORING SYSTEM: GENERAL REQUIREMENTS

All of the components of a monitoring system listed in the introduction could be onboard and operated in real time, particularly if immediate response to measured events is required (such as providing information to a ship commander on the impact on the ship structure of the current operations re, for example, sea state, ship speed, and heading). At a minimum, the first four components of a monitoring system, in the list in the introduction, need to be permanently installed onboard with the possibility of the other components being performed later in a more substantial facility off-board. The general requirements for the onboard components of the monitoring system include:

- Long endurance when battery power is required, particularly relevant for aircraft, since corrosion may occur during periods when the airframe is powered down.
- Light weight and small size: Adding extra weight is always an issue on aircraft, and the weight associated with large numbers of sensors on a ship may also start to become a problem. The volume associated with the sensor and interface may also be an issue in applications in confined spaces.
- Multipoint sensor networks to reduce network wiring, weight and complexity. A natural consequence of this is that communications must be digital, which in turn logically leads to the inclusion of a microcontroller on the sensor interface. The microcontroller could also be used to implement local data processing and communications with sensors that provide digital data.
- Local memory to store significant amounts of data prior to transmission. This can reduce the power requirements and the complexity of the communications protocol, particularly if a number of relatively high-speed sensors are located on the one network.
- Adaptability to accommodate new sensors: New and improved sensors are being developed every year.
- Remote software upgrade ability for "smart" sensors, to avoid the need to access the sensors as data-processing algorithms are developed and improved.
- The ability to handle large networks of sensors, since networking can significantly reduce the wiring requirements.
- The ability to time-synchronize sensor data collected over the whole sensor network.
- A single, consistent communications interface that is independent of the actual sensor.
- An architecture that allows simple or advanced data processing so that some of the data processing load can be moved to the sensor to reduce data communications and storage requirements.
- Electrostatic discharge (ESD) protection for increased robustness.
- Robust and long-life sensors.

An immediate consequence of these requirements is that the sensor interface will almost certainly require some local digital processing capability. This is most easily achieved by incorporating a low-power microcontroller into the interface, which immediately opens a number of possibilities for on-sensor data handling and processing.

10.4 MONITORING SYSTEM: HARDWARE

10.4.1 GENERAL CONSIDERATIONS

The electronics used for the sensor interface (SI) must be compact and lightweight, and this can be achieved by the use of standard surface-mount components and multilayer printed circuit board (PCB) technology. An SI board developed by the DSTO is shown in Figure 10.2. This consists of a core hardware (and software) block that provides a common interface for all of the DSTO sensors, and a specified hardware (and software) block to interface to two miniature, two-electrode electrochemical sensors, constructed on a four-layer PCB measuring 47 × 34 mm [5]. This SI has no connectors, since it is designed for permanent wiring and installation and weighs just 7.9 g. Figure 10.3 shows a schematic of the functional blocks for this DSTO SI. In this example, the SI connects to a multipoint four-wire bus that supplies power (4–16 V DC) and RS485 (TIA/EIA-485) communications, both with ESD protection. The RS485 serial communications standard was chosen because it is half duplex (thus only needs two wires), uses balanced data transmission (differential signals) that gives good noise rejection, has reasonable data rates, and has the capability to drive relatively long communications lines (up to 1.2 km). The latter requirement is particularly relevant for applications that may have networks over 100 m long in large ships. Other balanced-line interfaces such as M-LVDS (TIA/EIA-899) could be used for higher-speed communications, or the CAN bus (ISO11898), which is specified for shorter (40 m) networks.

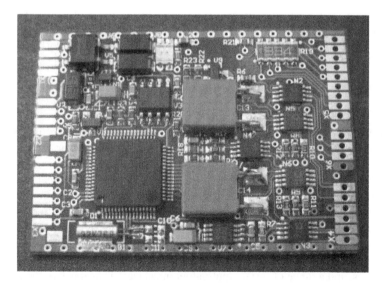

FIGURE 10.2 An example of an SI constructed on a four-layer PCB and measuring 47 × 34 mm with a total weight of 7.9 g. This SI is designed to interface with two miniature electrochemical electrodes and can measure resistance in the range 10^3–10^9 W, has an I^2C-compatible output, and has eight digital I/O lines for simple on/off measurements [5].

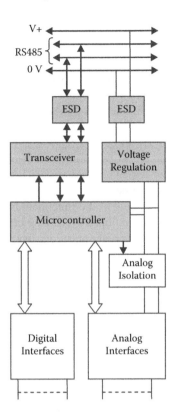

FIGURE 10.3   Schematic of the DSTO SI electronics illustrating the functional blocks and modular approach to circuit development. The upper, shaded region is the core hardware, common to all SIs. The lower, unshaded region is specific to the sensors that are interfaced with the SI.

10.4.2 MICROCONTROLLER

The use of a microcontroller on the SI greatly simplifies the interfacing with the communications bus and the hardware for interfacing with the sensor. More complex measurements can be performed, such as those that require the measurement of a dependent variable as another variable is changed, e.g., electrochemical impedance measurements that require a set of current measurements as an excitation voltage is changed. Data processing can also be used to lessen the communications and data storage requirements, e.g., not transmitting null or very low signals. Modern microcontrollers are low-power devices with inbuilt analog capability, such as analog-to-digital and digital-to-analog converters, pulse-width modulation, extensive timing capabilities, and a range of serial input/output interfaces. Thus much of the electronics required to interface with many sensors is already included in the microcontroller, reducing the amount of circuitry that needs to be included on the PCB. The microcontroller used must be crystal-timed to generate reasonably accurate time periods. It must also have internal counters and timers (preferably with a number of operating modes), a low-power/sleep capability, a reasonable amount of internal RAM (at least 2 K), and a high-speed serial interface capability, and it must be able to cope with large amounts of traffic on the serial interface (i.e., be able to ignore data traffic from other devices on a shared communications bus so that there is little or no impact on its own operation).

The DSTO SI mentioned previously uses a Texas Instruments MSP430F169® microcontroller to implement this. This fulfills all of these requirements and, for the last requirement, has a special Address-Bit Multiprocessor Format (ABMF) [6] for received serial data. Only data with the address bit set is recognized as received data in this format, so the microcontroller can be set up to completely ignore any non-ABMF data streams from other sensors. This is critical for low operating frequency (and thus lower power) so that the microcontroller (a) does not spend large amounts of time processing irrelevant data streams, (b) can remain in a low-power mode while other devices are transmitting data, and (c) does not require large serial input buffers to avoid serial input buffer overflow problems.

There are a number of other possible methods that can be implemented to address individual sensors. These include detection of (a) an "idle-frame," (b) a "break" character, or (c) a programmed character that the serial communications hardware can be set up to recognize independently of the processor.

- Idle frame addressing detects a period (e.g., 10 bits long) when the serial input is high and the next byte is then taken as an address byte. This mode can be used to generate an interrupt followed by examination of the next received byte by the processor. However, any data transmissions must ensure that there are no breaks of greater than 10 bits in the data stream. As will be seen later, this may be difficult, since it may be a requirement that transmission of data is interrupted to ensure that measurements of a sensor are made at fixed time intervals.
- A "break" character corresponds to a period (e.g., 10 bits long) when the serial input is low. This can be used to generate an interrupt that then deals with processing of the following data as an address. In the MSP430F169, a break is detected after a stop bit is missed, so the break period must be at least 20 bits long.
- The programmed-character technique requires specialist hardware in the serial communications section of the microcontroller that can detect a match of incoming bytes with the programmed character. This would be especially useful if combined with the ABMF mode; however, the Texas Instruments MSP430F® series of microcontrollers do not have a pro-grammed-character mode.

10.4.3   POWER CONSIDERATIONS

10.4.3.1   Low-Power Operating Modes

Critical to achieving low power is the use of low-power microcontroller technology that incorporates a wake-on-demand protocol. The two demands that the micro-controller must respond to are (a) external instructions and (b) the requirement to

TABLE 10.1

Average Current Used by an MSP430F169 in Different Modes

MSP430F169 Mode	Power Supply = 3 V Current (µA)	Power Supply = 2.2 V Current (µA)
AM	340	225
LPM0	70	65
LPM2	17	11
LPM3	2	1
LPM4	0.1	0.1

Source: Data from the MSP430x1xx user's guide [6].

periodically take and record the sensor measurements. In the previously mentioned Texas Instruments ABMF format, the microcontroller can be in a low-power mode with just the serial communications input, a universal asynchronous receiver/transmitter (UART), and the crystal oscillator operating. All serial data is ignored unless the address bit [6] is set in the received data byte. When this occurs, the microcontroller becomes active and can process the received byte. If each SI has a unique address code, then the SIs in a network of SIs can be configured to ignore all data other than data that is addressed to them. The microcontroller also includes internal timers that can be used to generate an interrupt and move the microcontroller from low to high power, perform measurements, and return to a lower-power mode. Table 10.1 lists the power used by an MSP430F169 in active mode and a number of low-power modes. Low Power Mode 3 (LPM3) of the MSP430F169 keeps one of the clocks active, so this can be used to generate an internal interrupt to periodically perform measurements.

10.4.3.2 Low-Frequency Operation

The frequency of the crystal oscillator and CPU clock speed also affect power consumption, with higher speed operation resulting in higher power. The Texas Instruments MSP430F169 microcontroller used in the DSTO SI can be used with a variety of crystal oscillators (up to 8 MHz) or an internal digitally controlled oscillator and can also be operated in a number of low-power modes. If very low power is required, then the DSTO SI can be configured with a 32,768-Hz watch crystal, with the CPU clock derived from the internal oscillator. Such clock speeds would be suitable for low-data-rate sensors such as temperature, corrosion, or environment sensors. However, these clock speeds may be inadequate for high-speed sensors and the consequent high data rates, such as strain gauges or accelerometers. The more recent Texas Instruments MSP430® family of microcontrollers can be configured with crystal oscillators up to 24 MHz, but with an associated increase in power requirements [7].

10.4.3.3   Communications

Another critical component for low power consumption is the power involved in the communication of data. The DSTO SI discussed here will generally spend most of its time in receive mode so that it can respond promptly to any network controller requests. Integrated circuits (ICs) such as the MAX3471® are a suitable choice for low-power RS485 communications with a supply current of just 1.6 μA in receive mode and only 50 μA when in transmit mode. However, this IC is limited to a maximum data rate of 64 kbits per second (bps) which is roughly 6 kbytes per second (Bps). Low communication rate is sufficient for small networks and/or low-data-rate sensors, but it will not be fast enough for larger networks and/or high-data-rate sensors. Unfortunately, higher-rate RS485 transceiver ICs tend to have higher operating currents. For instance, the MAX348xE® family of devices operates at 250 kbps, 2.5 Mbps, and 12 Mbps and, as a bonus, unlike the lower-power IC mentioned previously, these have built-in ESD protection. However, they have a supply current of 1mA in receive mode, with a slightly higher current in transmit mode. If power use is a particular issue, the RS485 transceiver IC could be placed in shutdown mode (with a current of 2 nA), with the SI periodically activating to check for activity on the network. An alternative would be to implement a DC signaling system so that a network controller could remotely enable or disable the sensor interface circuitry.

The actual communication rate required will depend primarily on the number of sensors on a network and the measurement rate. For instance, if the overheads involved in retrieving the data from the sensors (covered in Section 10.6) are ignored, then a single sensor making 100 2-byte measurements per second will produce 200 bytes of data per second. A serial communications rate of 2000 bps is just sufficient to continually transmit this as binary data. The protocol for the DSTO SI only allows printable characters, so that the data rate in this case is doubled, requiring at least 4000 bps. Thus at most 16 sensors producing data at the rate of 200 Bps could be operated on the one network using the low-power communications chip discussed previously at 64 kbps. Also, as discussed previously, there is a power-reduction advantage to operating at low data rates. Yet another benefit of low data rates is the ease of setting up the network and the higher tolerance to transmission line mismatches, such as those that could be introduced by poor or degraded connections. Lower-speed transceiver ICs are also often designed to be slew-rate limited to reduce transmission-line mismatch reflections and further simplify their use.

To enable use in both very low-power and high-data-rate applications, the DSTO SI design includes footprints for both high-speed and low-speed RS485 transceiver ICs and discrete ESD protection. Thus, any PCB can be loaded with the appropriate components for a particular application. Table 10.2 shows the current drawn by the DSTO SI for two crystal and baud-rate variants in different modes of operation. The quiescent current was calculated from the DC voltage measured across a low-value resistor in series with the positive supply to the DSTO SI. The current while the SI was transmitting was measured by monitoring the voltage waveform across a larger-value resistor in series with the positive supply with a differential-input digital storage oscilloscope. (A higher resistance was required due to the lower sensitivity of the oscilloscope compared to the digital voltmeter used for the static measurements.)

TABLE 10.2

Average Current Used by the DSTO SI (MSP430F169) in Active and Different Low-Power Modes

MSP430F169 Mode	High-Power RS485 Transceiver (7,372,800-Hz crystal oscillator and 115,200 Baud)	Low-Power RS485 Transceiver (32,768-Hz crystal oscillator and 4,800 Baud)
Active, receiving data	3.69 mA	339 μA
Active, transmitting data	8.20 mA (4.20 mA)	3600 μA (3400 μA)
LPM0	1.04 mA	48 μA
LPM1	1.04 mA	45 μA
LPM2	1.00 mA	10.7 μA
LPM3	1.00 mA	3.3 μA

Note: The values in parentheses are the approximate average currents in the 820-Ω RS485 termination resistor.

The average transmission current was determined by roughly integrating the area under the observed waveforms. In the case of the low-power transceiver, the transmission current was nearly constant; however, for the high-power transceiver, the current peaked at each data transition (i.e., a 1 to 0 or 0 to 1 in the data stream) and decayed quite rapidly until the next data transition. Peaks and troughs of around 14.8 mA and 3.7 mA, respectively, were observed, but with a bias to the low end. The RS485 line was terminated with an 820-Ω resistor, and the voltage generated over this was also monitored with the oscilloscope. The results in Table 10.2 show that in the quiescent condition, when transmission is not active and the MSP430 is in a low-power mode (LPM0, LPM1, LPM2, LPM3), the high-power transceiver and high clock frequency DSTO SI draws around 1 mA more current than the low-power version.

10.4.4 ON-SENSOR MEMORY

As indicated in the previous paragraph, a considerable amount of power use is associated with the network communications interface. Thus if power is an issue, it is desirable to communicate in bursts, with the communication interface disabled between data transmissions. This highlights the need for enough memory on the sensor interface so that measurements can be stored locally and transmitted in blocks. The amount of memory required will depend on the sensor measurement rate and the time between blocks of transmitted data. The provision of a local data buffer also makes it easier to manage large numbers of sensors on a network, since each sensor does not need to be accessed at a fixed time period; sensors with a local data store are tolerant of interruptions and delays in the network communications.

For low-data-rate sensors, modern microcontrollers may have sufficient internal memory to store many seconds to minutes worth of data (depending on the data rate). Higher-data-rate sensors will require the implementation of external memory. A DSTO SI (Figure 10.4) developed to interface with three-axis MEMS

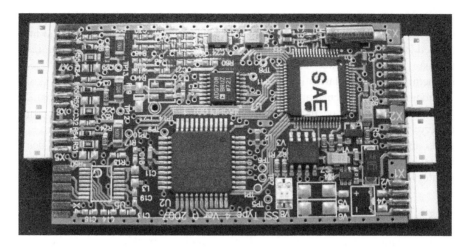

FIGURE 10.4   Example of a DSTO SI that includes 2 MB of memory for storing the measurements from a three-axis accelerometer.

(micro-electrical-mechanical systems) accelerometers has 2 MB of local random access memory and can store just under half an hour's worth of data locally for 16-bit measurements of three axes at 200 Hz. The data store is implemented as a First In First Out (FIFO) buffer controlled by a complex programmable logic device (CPLD). Even higher-data-rate sensors, or the ability to operate independently of the network controller for long periods of time, may require a much larger local data store for buffering or data reduction processing. To this end, the DSTO SI has been implemented with the option to include a micro Secure Digital Card® (SDC) connector on the underneath side of the PCB, beneath the core circuitry. SDC was chosen, since a simple serial peripheral interface (SPI) can be used, and the MSP430F169 has two universal asynchronous receiver-transmitter (UART) serial interfaces. One UART is dedicated to the RS485 bus communications, and the hardware has been organized so that the other can be used in SPI mode. A standard output pin is used as the chip select for the SDC so that the SPI bus can be expanded for other SPI devices if needed. Current SDC technology allows up to 32 GB of data storage, and this is expected to increase as the technology improves.

10.5   MONITORING SYSTEM: SOFTWARE

Any implementation of an SI requires software to interface with the network and software to perform and store the measurements and to possibly process the data. The network interface software should provide a consistent interface with the network, regardless of the sensor. In the DSTO SI there is an invariant core operating system (CoreOS) as well as user-defined software to manage the sensor-interface hardware and process sensor data [8]. The CoreOS oversees communications, power management, memory management, installation and execution of specific sensor software, and data-logging functions. A number of useful management routines have been developed by DSTO and can be downloaded by the user into the microcontroller's flash ROM. The CoreOS currently occupies only 24% of the available 48 KB ROM of an MSP430F169, leaving significant space for sensor interfacing code and any desired local data processing. An example of a more complex data processing task that has been implemented on the sensor interface itself is the performance of electrochemical, linear polarization resistance measurements [5]. This involves taking 128 measurements of current with varying voltage and then performing a (floating point) linear least-squares fit to these measurements to determine the resistance and the error in the fit. Thus, instead of transmitting all 128 measurements to the central controller, this DSTO SI only transmits two results. The MSP430F169 has been available since before 2005, and more recent microcontrollers have significantly greater ROM and RAM capacity to perform even more complex tasks.

10.5.1 COMMUNICATION PROTOCOLS

Communications with the SI should be kept as simple as possible, to reduce the associated overheads, but allow the user to implement complex procedures. To this end, an extensible command language with a set of simple, core commands allows the user to customize the sensors' response as required. Thus the level of complexity can be chosen to match the task being addressed as opposed to standards such as IEEE 1451 [9], which have a relatively high implementation overhead. Some form of error correction or detection should be included in the data stream so that the integrity of the data can be determined. Any number of data formats are possible, such as binary data or data with embedded tags. Some method to detect individual SI failure, such as a timeout if an SI fails to respond within a certain time, must also be implemented to avoid a single failure collapsing the whole network.

The DSTO SI implements a 2-byte cyclic redundancy check (CRC) and, since the system is to some extent experimental, all data are transmitted as printable characters, such as plain text or hexadecimal. A critical aspect of the DSTO SI communications is the use of the ABMF mode mentioned previously to lessen the SI overheads in monitoring and responding to network traffic. A simple timeout is implemented in the network controller, with the error being logged in the stored data.

10.5.2 IMPLEMENTATION ON THE DSTO SI

The software implementation on the DSTO SI will be referred to as an example to illustrate the requirements for low-power, extensive sensor networks, and so it is briefly discussed here. More detail is presented by Wilson and Vincent [8]. The DSTO SI has been developed with a view to keeping the hardware and software requirements low, particularly with regard to operating frequency (since higher frequency leads to higher power), but aiming to retain the capability of high sensor data rates. While some aspects are specific to the DSTO SI, any networked SI would need to provide the same general capability.

10.5.2.1 Addressing SIs: DSTO SI Groups

The DSTO SIs on a network can be organized into logical groups. Each SI always belongs to a unique group, so that it can be uniquely selected, and can be assigned to any number of other user-defined groups. An SI can only be active and accessed if one of the groups that it belongs to has been selected. The group-select command is the "|" character (pipe) transmitted in ABMF mode followed by the group name. If an active SI receives the group-select command and it is not a member of the group following the command, it becomes inactive.

Combined with the various operating modes of the MSP430F169 on the SI, the group structure provides a means for managing the overall power of a sensor system, since an SI can be in an inactive (low power) state and only become active when an ABMF byte is received. In this way, the SI can be set up to ignore all data traffic and not waste processing time or power receiving (and responding to) all of the data on the network bus. The group structure also provides a flexible method to determine which sensors will respond to the network controller. If a number of sensors share a group, then they can be selected simultaneously and would simultaneously respond to further commands (known as *tags*) from the controller.

10.5.2.2 Scripts and Tags

Scripts are executable code associated with a *tag*. Upon receipt of a particular tag, an active SI will execute the associated script. The scripts and tags are downloaded into flash ROM in the SI by the user under control of the SI's CoreOS. Thus the user can program the microcontroller to perform any desired actions with regard to sensor measurement and data processing. An added feature is the ability to download lists of tags and associate this with a single tag so that upon receipt of the single tag, the tags in the list will be sequentially executed. The use of these tag lists aids software development, since complex functions can now be easily broken up and tested in functional blocks. It must be noted that the location in ROM of a downloaded script is not known when the script is compiled; thus position-independent code (PIC) must be generated for these routines. Tags can have data associated with them to be read by the script.

10.5.2.3 Core Functionality

The minimum functionality for an extensible SI on a central, command-driven network includes the following:

- A unique name or address so that the SI can be individually accessed. This is provided by the unique group.
- A global name that all SIs share so that they can all be made active at the same time, combined with
- An inbuilt routine to respond to a network "explore" command that can be issued by the central controller to determine which SIs are on the network. Each SI responds to the explore command after a period of time determined by its unique name (to avoid communication overlaps).
- An inbuilt routine to download, store, and manage executable code (a script) and associate this with a command (a tag).

With these capabilities, the user could then download all of the scripts and tags that are required for operation of the SI. The DSTO SI implementation includes the previous capabilities together with a few other built-in scripts that make operation of the system easier, particularly during software development. These include:

- A simple acknowledge script that is useful to detect that an SI is active.
- Deletion of previously downloaded scripts, tags, and groups. This is effectively a small file management system. The tags and scripts are stored in one linked list, with the groups in another linked list in ROM.
- Enabling and disabling the flashing of ROM—a safety feature to avoid inadvertent modifications of flash ROM.
- Downloading tag lists.
- Loading and reading a unique string for an SI. This enables methods to automatically manage the types of SIs attached to a network. A small amount of flash ROM is set aside to store this information.
- Control of the power modes that the SI reverts to between measurements.
- Setting up and moving into a data-logging mode (described later in this section).

As well as these scripts, it is useful for the CoreOS to expose a table of entry points for various commonly used utility routines within the CoreOS so that user scripts can access them. Such routines include communications functions, methods to locate tags and scripts so that a script can call other scripts, methods to access the SDC if it is included and, for instance, access to shared lengthy code for (say) floating point operations.

It is also desirable to implement an interrupt jump table in RAM so that interrupt routines can be dynamically changed by altering the vector in RAM. In the MSP430F169, the interrupt vectors are stored in ROM, so this is implemented by the use of dummy interrupt routines that then execute a call to the vector stored in the RAM table. More recent versions of the MSP430Fxxx microcontrollers have the capability to define a secondary interrupt vector table, so there is no need to use the dummy routines, and the execution of the interrupt will be faster. Making the interrupt table accessible is desirable, since user routines may wish to implement interrupt-driven processes. Access to the interrupt table is essential for the data-logging activity as implemented in the DSTO SI, since it is driven by an internally generated timer interrupt that executes a user-downloaded measurement routine (see Section 10.6.2, SI Data-Logging Measurements).

A further desirable function is to have the ability to execute a start-up script that is executed when the SI is initially powered. This script should execute after all standard start-up initialization has been performed so that the user has complete control over how the SI starts. This could be used, for instance, to automatically recover from a power-down so that data collection continues automatically after the power is reapplied. The DSTO SI implements this with a script with the name "auto." If the CoreOS detects the presence of the "auto" script upon start-up, it automatically executes the code.

10.5.3 ABMF MODE

Adoption of the Texas Instruments ABMF mode, while extremely useful with regard to communications, power control, and selection of specific SIs on a network, has some limitations in implementation with standard serial communications. Standard serial communications protocols do not implement ABMF mode directly, so a work-around is required. The ABMF bit occurs at the location that the parity bit occurs in standard serial communications, and thus it can be implemented by setting the serial parity bit to a mark in the transmitted data stream and as a space when ABMF is not used. PC communications with the DSTO SI has either been through the standard RS232 port with an RS233-to-RS485 converter, via a USB port using a USB-to-RS485 converter or via ports on an RS485 IO card. It has been discovered that there can be significant latency in setting and resetting the serial parity bit (necessary to go into and out of the ABMF mode). It is suspected that the particular PC driver software implementations are the cause of this latency, although this has not been confirmed. The internal RS232 driver gave the least latency, and so for larger networks or higher data rates, where the latency impacts on data throughput, the RS232-to-RS485 converter approach is preferred if the network controller is an embedded PC or similar commercially available device. A longer-term solution for this issue would be to write a driver specifically for this application with a special ABMF bit-set command. Another alternative is to use an "escape code" or "modem sequence" approach and an MSP430F with two UARTs to receive and process the data and escape codes, and to then relay the data to the DSTO SI network using the native ABMF. A further approach is to use MSP430F devices as the network controllers and data loggers and thus avoid the problem.

10.6   MEASUREMENT METHODS

The critical consideration for any measurement and data logging is that the time required to receive data from all of the sensors on a network is less than the time that all of the sensors took to acquire the data. If this condition is satisfied, then the network will be able to retrieve and record the data faster than the measurements are occurring. For example, 115,200 Baud corresponds to roughly 10,000 Bps. Thus, if all of the sensors on a single 115,200-Baud network generate less than 10,000 bytes of data per second, less the number of command bytes required to initiate data transmission, plus any other overheads, then the data will be read from the sensors faster than it is accumulated.

The measurement and retrieval of data in a network of sensors can be performed in a number of ways:

- The SIs are individually polled by a central controller/data store to make and transmit measurements.
- Measurements are initiated by the SI and transmitted by the SI to a data store.
- Measurements are initiated and stored by the SI, but data transmission is under the control of a central controller/data store.

The second method will not be considered, since it raises communication complexities associated with detection and dealing with data collisions and/or signaling for control of the bus. The other two methods are command driven with respect to activity on the network, with the central data store determining what traffic is on the bus, and thus there will be no data-collision issues.

10.6.1   SI Polled Measurements

With polled measurements, an external network controller sends a request for a measurement and waits for the results to be transmitted back to it. The sequence of events would be to select the required SI, send the command that initiates the measurement and transmits the results, and wait for the data to be transmitted. The SI would then be deselected after receipt of the data by the controller, and the next SI selected. (Selection, deselection, measurement, and transmission could all be initiated with a single command to reduce the communications overheads.) The communications bus will not be available for any other activity while the measurement is being made and transmitted, and thus this method is not recommended for sensors that may take a significant time to perform a measurement. Another complexity is that the time at which measurements are made will be different for the different SIs, and thus will not be suitable for high-speed sensors, where the correlation of the signals may be important (e.g., for structural load and response measurements). Polled measurements for high-data-rate sensors will also pose a challenge, since there can be a significant overhead to measure and retrieve the data because a command requesting the data must be transmitted for each data point. Using a polled measurement approach severely limits the speed and number of sensors that could be deployed on one network.

The DSTO SI tag, script, and group structure can be used to reduce some of these drawbacks. For instance, two tags could be used, one to initiate the measurement and one to request transmission of the measurement(s). The measurement tag would be sent first and, after an appropriate time lag, the transmit tag would be sent. Other bus activities could occur during the time between the two tags, such as initiating measurements on other sensors. Note also that with the DSTO SI protocol, a single group select followed by a single measurement tag could be used to initiate measurements on a number of SIs at the same time, with the data retrieved using different transmit tags for each SI. This would ensure time-synchronized measurements. Even with these methods, a better approach in many instances is to use a data-logging method, as discussed in the next section.

10.6.2   SI Data-Logging Measurements

In this case, the SI itself has the capability to periodically perform measurements and store the data locally with no intervention from an external controller, apart from this data-logging action being initiated. To do this, the SI must have a sufficiently large memory to store a number of measurements and an accurate time source, such as a crystal oscillator, so that these measurements are performed at precise times. The SI must also respond to requests from the external controller to transmit the stored data and perform other functions such as time synchronization.

The actions to perform the measurement, data transmission, and other functions are significantly more complex than the simpler polled approach discussed previously. This is because the measurement and network activity are asynchronous. Measurements may be occurring when a command, possibly for the SI, is issued by the controller, or the SI may be responding to such a command when it is time to perform a measurement. Thus, a number of possible interactions must be considered, and the relative importance of the particular actions (e.g., responding to the controller versus taking measurements at a fixed period) will determine exactly how data logging is implemented.

10.6.2.1 Measurement and Transmission Routine Interrupts

For moderate network bus speeds, it is highly likely that for high-measurement-rate sensors, the transmission time will take longer than the time between measurements. Thus, to maintain the integrity of the measurement timing, the data transmission must be periodically interrupted as measurements need to be performed. For instance, if 16-bit data are being recorded at 300 Hz (i.e., every 3.3 ms) and the results transmitted every 10 s, this corresponds to 6,000 bytes of data to be transmitted. At 115,200 Baud (approximately 0.1 ms per byte), this will take around 600 ms, and in this time, another $600/3.3 = 181$ measurements will be made. A related complication is that the continual interruption to perform the measurements could add up to a large transmission overhead if the measurement takes a significant time. Using this example, a measurement that took 2 ms to perform could result in more than 360 ms (2 ms × 181 measurements) added to the transmission time. This might also need to be avoided, and to do so, the transmission routine would also have to be able to interrupt the measurement routine to send the next byte of data as soon as possible.

Most UARTs on microcontrollers have at least a single transmit buffer memory and generate an interrupt when this buffer is transferred to the serial output register. Thus if the microcontroller can respond to load a new byte into the buffer within the time that it takes to send a byte, then the data will be transmitted in a continuous stream. At 115,200 Baud, this time is around 0.1 ms and corresponds to 700 clock cycles for a microcontroller operating at 7 MHz, more than enough to implement an interrupt routine that simply loads a byte into the buffer. However, allowing the transmission routine to interrupt the measurement routine will inevitably lead to an unwanted increase in the variability in when the measurements are made. This variability can be reduced in many modern microcontrollers that support direct memory access (DMA), which could be used to transfer data into the UART, with no processor intervention required after the transfer has been initialized. In this case, the processor would normally be halted while the DMA unit accessed the memory, but this time would be short compared to responding to an interrupt. The DMA function is, however, generally limited to the internal memory of the microcontroller, so it can not be used with external memory.

Figure 10.5 depicts a simplified state diagram showing how the measurement and transmission processes may interact so that the two activities are interleaved. The exact details of the software required to perform these functions must be carefully managed to ensure the integrity of the measurements and the transmitted data and any other commands that might be implemented, such as time-synchronization actions. If there are time-critical parts of the measurement routine, then interrupts would have to be disabled while these were being performed. In any particular application, this may place a limit on the maximum speed at which a sensor could be operated. The ability of the transmission routine to interrupt the measurement routine introduces a certain amount of uncertainty in the actual measurement time.

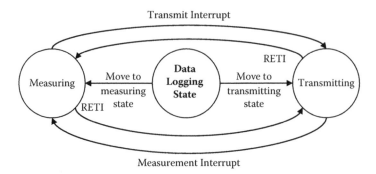

FIGURE 10.5 Simplified state diagram showing the interrupt capability between the measurement and data transmission routines. *RETI* indicates return from the interrupt.

With the MSP430F169 used in the DSTO SI operating at around 7 MHz, this could correspond to a few microseconds. For most applications, this degree of timing inaccuracy will not be significant and, as discussed later in greater detail, the time to receive and process commands from the network controller is much longer and thus more critical. If needed, very-high-speed systems could be implemented using a double-buffering approach with multiple microcontrollers, one purely for communication and the other purely for performing the measurements.

As well as initiating data transmission, another particular requirement will generally be the need to be able to time-synchronize the measurements made by different SIs. For instance, each DSTO SI has its own crystal oscillator, so that over short periods of time, the clock difference between different SIs will be small. However, over longer times, the different SIs will lose synchronization, and thus the SIs must be able to receive and process time-synchronization signals.

10.6.2.2　Communications Interrupts

The SI must be able to respond to commands issued by the controller to initiate the data transmission and perform any other actions. To ensure that these commands are not missed, the SI must respond and process any received network data before another data byte is received (within 0.2 ms at 115,200 Baud). The time to process the received data will introduce more uncertainty in the time at which measurements are performed, and this is likely to be greater than the effect due to the data-transmission interrupts discussed previously. An alternative would be to simply store the received command bytes in a local buffer that is checked after the measurement routine finishes. The time to do this would be quite short (a few microseconds), but it runs the risk of buffer overruns if too many characters are received during the measurement time. This could be avoided by restricting the maximum length of any command, with any further characters (e.g., due to a different SI placing data on the network bus in response to a request to send data) being ignored.

10.6.2.3　DSTO SI Implementation Example

Data logging on the DSTO SI has scripts defined to perform the following data-logging actions plus a special command to move the SI into data-logging mode [8]:

- *Measurement.* The script that is executed whenever the data-logging period has passed. The period is determined using an internal timer with an interrupt capability.
- *Transmit.* The script to transmit blocks of measurements to the network controller, or to initialize this transmission of data, whenever the associated transmit tag is received.
- *Time stamp.* The script that processes the information (an identification code) received with the time-stamp tag from the network controller and stores this information, along with appropriate local timing information, in the local data stream. Thus, the clocks of the different SIs are not physically synchronized, but there is sufficient information in the data stream to determine the time differences between different SIs on a network.
- *Stop.* The script that exits the data-logging mode.
- *Initialize.* The script that is executed as the last action in setting up data-logging mode and can be used to initialize data structures or perform any other required actions.

All of these scripts are defined and downloaded by the user, and thus the specific actions are determined by the user. The command that sets up a DSTO SI for the data logging also includes information on the logging period, the amount of RAM that is required (optional), the maximum number of data values to transmit at a time, the type of measurement being performed (slow or fast, see later discussion), and the uppercase alphabetic character (the *data-logging character*) that is used to recognize commands for this particular SI when in data-logging mode. In the data-logging mode, the transmit, time stamp, and stop tags are received in ABMF mode (i.e., the first character must have the address bit set), and the first character must be the data-logging character. Use of this special addressing means that data-logging SIs ignore all other serial data, and thus the measurement activities on an SI are not affected by the bus traffic. The constraints for the DSTO SI data-logging actions are as follows:

- Timer A in the MSP430F169 is used to generate an interrupt at the period defined during data-logging setup. When the user-defined number of interrupts is reached, the measurement script is executed. Thus Timer A can not be modified for use with any other functions.
- Measurements, or any other actions, must not disable interrupts for greater than the time it takes for a single character to be received on the RS485 input to ensure that the data-logging character, and any subsequent characters, are not missed. At 115,200 Baud, this corresponds to around 0.1 ms.
- Receipt of a character in ABMF mode must always be serviced and checked for receipt of the correct data-logging character. If this character is correct, then further normal data characters must be received and either acted on or stored.
- SIs must respond promptly to requests to transmit data so that the network is not held up for long times.
- SIs must always be able to receive and store time-stamp information and embed this in the SI's recorded data.

10.6.3　SI Response and Timing Considerations During Data Logging

A strategy has been adopted so that taking measurements and responding to data-logging requests from the central controller (such as a request to transmit data or requests to store time-stamp information) do not interact adversely with each other. This strategy depends on the user identifying the speed of the measurements made for the sensors, which are considered to be either fast or slow sensors. Examples of fast measurements would include sensors for strain and acceleration that may be sampled at hundreds of hertz and take less than a millisecond to obtain the measurement. Slow measurement would include corrosion sensors that can take many minutes to perform a single measurement. Other sensors, such as temperature sensors, may be sampled with long time periods between measurements but still perform any specific measurement very rapidly, and these would thus be considered to be fast sensors. Measurements may also have time-critical parts that need to occur without interruption. Thus the speed at which a measurement is performed and whether a measurement has time-critical parts are important considerations when the microcontroller routines for the scripts are being written. To this end, in the setup phase for the DSTO SIs, the user defines whether measurements are fast or slow, and the CoreOS stores this information. The CoreOS sets flags so that if a measurement is in progress, the measurement routine can check to see whether it has been interrupted by the receipt and/or response to a command. The measurement routine can then take the appropriate action (e.g., restart the measurement if it was interrupted in a long time-critical part of the measurement). The measurement routines must obey the requirements set out in the following subsections in addition to the previously discussed requirements.

10.6.3.1　Slow Measurements

Slow measurements must always be able to be interrupted to respond in a timely manner to requests for transmission of data and to process received commands such as the time-stamp command. (In the DSTO SI, only an ABMF byte with the correct data-logging character is responded to.) As a consequence of the need to process and respond to commands, a slow measurement may end up being delayed or interrupted by at least the time taken to transmit a block of data or respond to the time-stamp command. Since these measurements are (necessarily) performed at large time intervals, the actual time that a slow measurement is performed is generally not critical, and differences of a few tens to hundreds of milliseconds will not be significant. In the worst case for a measurement that must be made without interruption, if it was interrupted just near the end of a measurement, then it would have to be repeated, leading to a delay of a full measurement time. If this measurement timing skew is unacceptable, then the measurement routine could also record in the data stream any delay in the time that the measurement was actually made, i.e., it would record the actual measurement time. Another critical consideration is that the time between requests for data must be greater than the time to make a measurement; otherwise, it would be possible to reach a situation were a measurement is never completed due to interruptions to transmit data. This requirement is a practical requirement anyway, since it would not be productive to ask for data on a timescale shorter than the sensor measurement period.

10.6.3.2 Fast Measurements

Fast measurements are always able to interrupt any other processes to make a measurement to ensure that the period between measurements will not vary too much. However, any critical measurement times (i.e., measurement times that can not be interrupted and would thus briefly disable interrupts) must be brief to (a) avoid large gaps in data transmission and (b) ensure that all valid data bytes for commands transmitted to the SI are received. Thus, at 115,200 Baud, the maximum noninterruptible period is 0.1 ms. Commands (with an ABMF byte with the correct data-logging character for the DSTO SI) are always able to be received and stored, but if a measurement is in progress when the last byte of the command is received, the command will not be acted on until the measurement finishes. Thus the initial response to a transmission request or the processing of a time stamp may be delayed by the time that it takes to make a measurement. In order to reduce time slew for the time stamp, the local time when the data-logging character of a command is received is always recorded so that even if there is a delay waiting for a measurement to finish, the time that the command was received is correctly recorded.

10.6.3.3 Implementation of Data Logging

As discussed previously, and reiterated here, it is most desirable that transmission of data from the SI be performed using an interrupt routine that is itself interruptible by the measurement routine for fast measurements. Thus the two actions of transmitting data and performing the measurements may interleave with possible brief delays in either or both of them due to processing of the other routine.

A flow diagram of the data-logging state and the routines involved in this process for the DSTO SI is shown in Figure 10.6 and discussed in the following subsections.

10.6.3.3.1 Main Routine: Internal to the CoreOS

The main routine is normally halted in a low-power mode or in a loop to check for received network commands data. Whenever an interrupt occurs, the main routine will always check for received command characters.

10.6.3.3.2 Command Receive Routine. Internal to the CoreOS

This interrupt-driven routine receives network data. The first interrupt requires the AB bit to be set (ABMF mode), and it is checked to see whether it is the data-logging character. If it is the data-logging character, then subsequent non-AB bit set characters are received and stored in a serial buffer. The internal SI time that a command is received is stored for time-stamp processing. The major difference is that for slow measurements, any received command data is processed straight away, whereas for fast measurements, the received bytes are stored and processed in the main routine. This is done so that if a command is received while a slow measurement is in progress, the command will be acted on immediately rather than having to wait for the measurement to finish (which might be minutes). If a fast measurement is in progress,

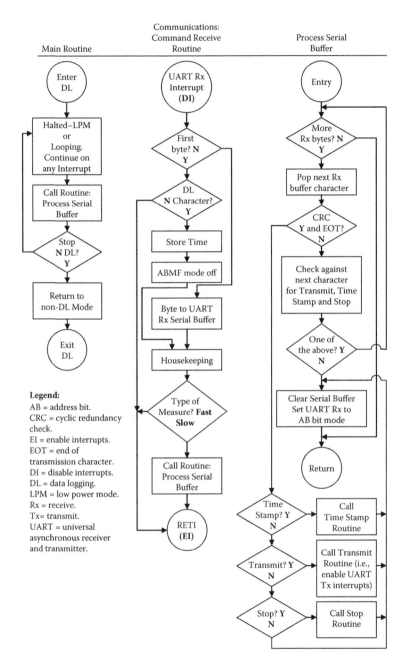

FIGURE 10.6A Flowchart showing the main data-logging routine, the response to a received character, and the processing of a received character when a DSTO SI is in data-logging mode. Received data is processed immediately for slow measurements, since if a measurement was being made when the data was received, the RETI would return the processor to the measurement routine, which could take a long time to finish, delaying the response to any received command.

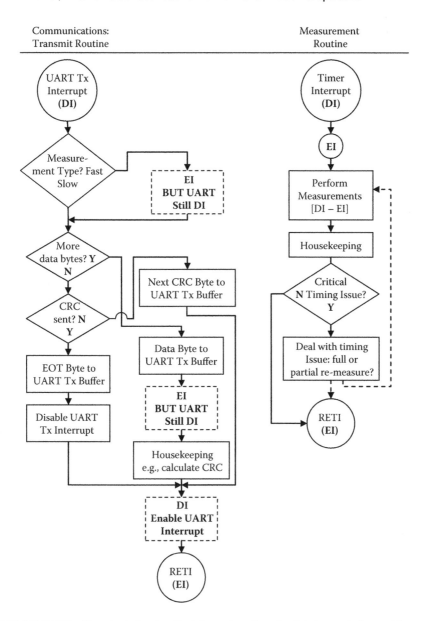

FIGURE 10.6B Flowchart showing the interrupt routines for data transmission and for performing timed measurements. Data transmission may be interrupted between bytes, but in this implementation, when the interrupt is being serviced to place the next byte in the Tx buffer, it may not be interrupted. The measurement routine is always interruptible, since the processor may need to service either received commands or requests to transmit data. The DI— EI indicates that short periods when interrupts are off are allowed. Dotted boxes indicate optional actions. The actions in brackets indicate an automatic microcontroller action.

the command must wait until this measurement finishes before it will be acted on. No other interrupts can occur while a byte is being transferred to the serial buffer.

10.6.3.3.3 Process Serial Buffer: Internal to the CoreOS

This routine checks the received command data firstly for whether it matches the tags in the SI (either transmit, time stamp, or stop tags) and then whether it has a correct CRC followed by an end-of-transmission (EOT) character. If these conditions are met, the appropriate data-logging routine is called. If at any point none of these conditions can be met, the serial buffer is cleared and the routine exits. This process is interruptible by fast measurements but not by slow measurements.

10.6.3.3.4 Transmit Routine: Sensor Specific, Provided by the User

All data transmission is best implemented as an interrupt-driven routine with a new byte requested whenever the transmit buffer is empty. This routine responds to the interrupt by loading in the next data byte or the CRC or EOT bytes. If the EOT byte is loaded, then the UART transmit (Tx) interrupt is disabled, since all data will have been sent when the EOT is sent. During data transmission, a CRC is calculated for all data bytes and appended to the data stream. Interrupts may be enabled on entry only for fast measurements to reduce measurement timing errors. This is only advisable for very quick measurements; otherwise there may be too much time penalty waiting for a measurement to be performed. An alternative is to enable interrupts after the buffer has been loaded and prior to any more lengthy calculations (such as calculating the CRC), since now the transmission and the calculation will occur in parallel. The UART Tx interrupt would need to be individually disabled to ensure that the interrupt routine is not called until it has finished. This is the point where the trade-off between fast response to the network or minimizing measurement timing errors must be made.

10.6.3.3.5 Measurement Routine: Sensor Specific, Provided by the User

The CoreOS automatically enables interrupts before transferring to the measurement routine. This routine does all sensor measurements and data storage and may temporarily disable interrupts, but this must be less that the time to receive a single command byte; otherwise these could be lost. The command-receive routine raises flags to indicate to a measurement that it has been interrupted. The measurement routine can ignore or respond to these flags, depending on the measurement process and whether this occurred during a critical part of a measurement. For the DSTO SI, the measurement routine also needs to be aware of whether a time-stamp command has been received and where the data associated with this will be inserted in the stored data.

10.6.3.3.6 Time Stamps: Sensor Specific, Provided by the User

The time-stamp routine is initiated by the process-serial-buffer routine and is interruptible. Essentially, it places the time-stamp information into the stored data. Exactly when this is done is not critical, since the time that the command was received is stored. However, the routine must coordinate with the measurement routine regarding where the data is stored.

10.6.3.3.7 Summary

The system described here can be used to implement data logging on a single microcontroller SI and reduce the timing errors that this will introduce to the measurement of the sensors. This is really only relevant for fast measurements. Most timing errors will occur when the command-receive routine is serviced either just before or during a measurement, so this routine needs to be carefully crafted. Examination of the machine code generated for the DSTO SI for the command-receive routine indicates that around 1,000 clock cycles are required for the longest processing route. This is associated with checking that the correct data-logging character has been received, changing the mode of the UART, and checking the data-logging type (fast or slow). At around 7 MHz, this corresponds to less than 0.2 ms, and this would be sufficient for most structural-monitoring applications. (Note that this processing time also limits the maximum Baud rate of this system to 115,200 Baud.) If closer timing was required, the actual measurement timer counts at the time that a measurement is performed could also be recorded. Since the measurement is triggered when this count reaches a certain value, the deviation from this value will give a measurement time accurate to within one period of the measurement timer counter, which is around 1 μs for the DSTO SI implementation considered here.

A hardware solution could involve the use of two microcontrollers: one dedicated to the sensor measurements and the other dedicated to the network communications. This is the only way to ensure that measurements will not be interrupted by the need for the processor to handle incoming commands unless a software handshaking approach was adopted. With software handshaking, the controller could send a single character in ABMF mode and then wait for the SI to respond before sending another single character. In this way, a command could be "clocked" in, character by character. The SI would only examine these when a measurement was not occurring. This approach would limit the number of sensors on a network, since each sensor would need a unique, single-character name. For binary data, this would be 256 sensors and fewer than this if data streams have some format limitation, e.g., limited to printable characters. However, this approach could also compromise time synchronization, since now the time at which a time-synchronization command was received could be delayed by up to the time it took to take and store a measurement.

10.6.4 SMARTER MEASUREMENT TECHNIQUES

The previous discussion has implicitly assumed that the sensors are continually measuring and recording, since this is the easiest way to implement a sensor interface. In many instances, the measurements may either be irrelevant (such as very low strains or accelerations) or not changing significantly over extended periods of time. In these cases, it would be smarter to periodically record an average signal or a "not changed" parameter rather than large strings of zeros. This can be achieved with a microcontroller-based SI by examining the data as it is measured and then taking the appropriate action. In some cases, the recording of pre-event data is desirable to give a baseline prior to a significant change in the measured signal. This can also be achieved with the appropriate software. Using this event-driven data-storage approach could significantly reduce the required data-storage capacity.

Adoption of this smart approach to data recording also reduces the average power consumption for an SI by decreasing both data transmission times and memory writing times for fast sensors that use RAM (or equivalent data stores) that are external to the low-power microcontroller.

10.7   EXAMPLE SENSOR INTERFACES

DSTO has, and is developing, a number of different DSTO SIs for a range of applications. These are all based on the DSTO SI detailed briefly in the previous discussion and have the same physical width and range from 47 to 60 mm in length. The sensor capabilities developed include:

- Three-axis strain-gauge rosettes at up to 200-Hz data rates per axis
- Two- or three-axis analog MEMS accelerometers at up to 200-Hz data rates per axis
- Three-axis digital MEMS accelerometers (using SPI) at up to 200-Hz data rates per axis
- Remote temperature monitoring
- Measurement of DC resistance from 10^3 to 10^{10} Ω with low excitation voltages (less than 20 mV, useful for electrochemical measurements) [5]
- Measurement of paint/sealant/bond degradation using thin wire sensors [10]
- Measurement of AC impedance (amplitude and phase) from close to DC to greater than 400 kHz
- Measurement of DC resistance from hundredths to hundreds of ohms (useful for mass-loss sensors)
- Corrosion sensor suite including $2 \times$ DC resistance (10^3 to 10^{10} Ω), temperature, humidity, and two closed/open circuit sensors [5]
- General measurement of current from picoamperes to milliamperes
- General measurement of voltages from millivolts to tens of volts
- Interfacing with COTS sensors that have current or voltage as an output signal
- Interfacing with COTS sensors that have digital serial outputs such as I²C, SPI, RS485, or RS232
- Interfacing with COTS devices that produce NMEA data streams (e.g., GPS devices)
- Digital on/off sensors

Since all of these examples have the same network interface and core operating system firmware in the DSTO SI, a comprehensive network can be easily put together to monitor for corrosion, structural loads, and other aspects of large operational platforms.

As an example, DSTO currently has a network of strain, acceleration, movement, corrosion, and temperature sensors installed on one of the Royal Australian Navy Armidale Class Patrol Boats [10, 11]. This consists of around 100 sensors located at 22 points throughout the boat. The controller, an embedded PC, is located centrally, so the network actually consists of a number of networks going fore and aft of the controller, all synchronized by the controller. This improves the robustness of the system, since faults on the different networks will be isolated from each other. Each network has a combination of fast and slow sensors, but the form of the interface for each sensor is the same, and they are all treated in the same way by the controller. The sensor initialization and data-logging sequences for the sensors are stored as text strings that contain the tags that are sent plus timeout information. Since the interfaces are consistent, the format of these text sequences is the same for all of the different sensors.

10.8   DATA MANAGEMENT

The networking system considered here has been for a network with a central controller and data store. One advantage of such a system is the speed of data retrieval, which essentially just involves swapping out the mass data storage device. An alternative is to have the data stored remotely (e.g., on the sensors) and then downloaded when the data are to be retrieved. This could take a considerable amount of time. For instance, over 3 months, 20 strain rosettes operating at 100 Hz make around 50×10^9 measurements, corresponding to 800 Gb of data. Even at a data rate of 100 Mbps, it takes 8,000 s to retrieve this. At 1 Mbps, the data download would take in excess of 9 days. There is also a required added overhead and complexity if the data are downloaded on retrieval, in that the sensor network must be able to operate at high data rates, whereas if the data are continually dumped in small blocks to a central data store, a much lower (and easier to implement) network bandwidth is required.

Currently, the easiest and cheapest way to archive the large amounts of data likely to be collected from an extensive structural management system is to use large-capacity hard discs. A 2-TB hard disc could store over 5 years of the strain-gauge data envisioned in the example presented here. Periodic copying of the data to a fresh disc would ensure data integrity.

10.9   INFORMATION FROM DATA

Sensors are not applied without a purpose, and the collection of large amounts of data is not useful unless these are related to the structure that they are monitoring. For structural sensing of strain and acceleration, this will often require correlating the observed measurements with theoretical values obtained from finite-element models. Depending on the level of detail involved, these models can be very complex and time consuming to establish, perform calculations with, and validate. Unfortunately, load-related issues (such as fatigue cracking) are often quite localized and depend intimately on local topology and thus require detailed modeling, at least at a local level. Other approaches involve detecting changes in the global distribution of loads, or other measurands, and relating these back to baseline measurements. These techniques offer the possibility of flagging a problem area that can then be investigated with either conventional NDE techniques or by implementing a locally dense sensor network within the whole sensor network.

Corrosion is a major cost in the maintenance of large assets, and much work has gone into the development of corrosion sensors and the interpretation of corrosion sensor data. The ultimate goal is to develop prognostic techniques based on sensor measurements. Again, corrosion tends to be a localized event, so the inclusion of sensors that directly detect corrosion at a particular location is of limited value. One approach to overcome this limitation is to model corrosion in corrosive environments in a probabilistic way [12, 13]. The measurements from a suite of environmental sensors are then used in these models to determine the probability of the occurrence of certain degrees of corrosion. Another approach is the inclusion of small, low-profile, and small-footprint sensors that can monitor extended regions [14].

In general, the interpretation of the sensor data as it relates to structural properties requires a detailed understanding of both the structure and the properties of the materials that make up the structure. In many instances, this will not be straightforward, with the application of sensors being the end result of a prolonged study, followed by data analysis, which will require a further major effort.

10.10 SUMMARY

The current state of technology for electronics and data storage is sufficiently advanced to implement reliable low-power sensor interfaces in extensive networks. Communication data rates up to millions of data points per second are attainable with sensor interfaces operating at powers in the range of tens of milliwatts. Lower data rates can be achieved with less than milliwatt power sensor interfaces. Future trends in the development of ICs will see power requirements reduced farther with increased digital and analog capability.

While wireless networks are attractive—since they require no cabling—the power requirements for higher-current sensors (such as strain gauges) and the concomitant requirement for manual battery replacement make a wired multipoint network the preferred implementation of a general, low-power, extensive network of sensors for structural monitoring. The sensor interfaces on such a network include a microcontroller to handle network communications and to implement smart-sensor capabilities, such as local data processing, event-triggered data collection, and local data store. The SI developed by DSTO is an example of a particular implementation for an extensive sensor network with low-power nodes.

Care needs to be taken in the development of the software on the sensor interface to ensure that large amounts of network bus data traffic does not impact on the operations of a networked sensor, particularly in regard to the timing of measurements. Also, methods must be implemented to allow simultaneous data transmission and measurement so that lower-speed (and thus lower-power) data-transmission rates may be used. The interaction with the sensor measurement routine of both the data transmission and the receipt of commands from the network needs to be carefully managed to ensure that large timing errors are not introduced with regard to the time that the measurements are made. Methods must also be implemented to synchronize the data on the separate sensors in a network. If high-speed measurements are to be made (in excess of 1,000 points per second), then it might be necessary to perform the functions of network communications and sensor interfacing in separate microcontrollers to ensure accurate timing of data collection. Microprocessors with much higher speeds than those considered here could also be considered, along with the resultant need for higher power.

An extensive sensor network is best managed by a central controller to handle data storage and synchronization of the whole network. Sensor interfaces that can operate as autonomous data loggers—with data being requested and transmitted in blocks—are recommended rather than having the central controller poll each sensor for each measurement. This improves the response time of the network, simplifies time synchronization, and reduces the power required for the whole network.

In the current state of the art, the biggest challenge for structural-monitoring datagathering systems is how to ensure the integrity and robustness of the sensors. The challenge for the application of structural monitoring is the correlation of sensor measurements with the state of the structure. This requires extensive modeling and knowledge of the structures and the materials properties of the components of the structure as well as the expected loads and operating environment.

ACKNOWLEDGMENTS

The author would like to acknowledge the continuing contribution, over many years, by Peter Vincent to the development of the DSTO SI. Also acknowledged is the work done by CPE Systems, under contract to DSTO, on the initial development of the DSTO SI core operating system software and for work on PCB layout and software development for a number of specific sensor interfaces. The contribution by Ian Powlesland to early discussions on the form of the DSTO SI is also acknowledged. Finally, the contributions of Andrew Finlay, Toby Seidel, Richard Muscat, Oscar Vargas and Ladislav Zeve to the design, construction, and testing of various forms of DSTO SIs are acknowledged. The continued development work and application of specific DSTO SIs has increased the author's depth of understanding of the practical requirements for sensor systems. Much of the work at DSTO was funded through DSTO's Corporate Enabling Research Program.

REFERENCES

1. Boller, C., F. K. Chang, and Y. Fujino, eds. 2009. *Encyclopaedia for structural health monitoring*. New York: Wiley.
2. Staszewski, W., C. Boller, and G. Tomlinson, eds. 2004. *Health monitoring of aerospace structures*. New York: Wiley.
3. Hu, Y., and B. G. Prusty. 2007. A new method for oil tanker structure condition monitoring. *Ships and Offshore Structures* 2 (4): 371–77.
4. Cooke, G., G. Cooke, and G. Kawanishi. 1998. *A study to determine the annual cost of corrosion maintenance for weapons systems in the USAF*. Prepared for AFRL/MLS-OL by NCI Information Systems, Inc., Contract No. #F09603-95-D-0053, February.
5. Wilson, A., P. Vincent, P. McMahon, R. Muscat, J. Hayes, M. Solomon, R. Barber, and A. McConnell. 2008. A small, low-power, networked corrosion sensor suite. *Materials Forum* 33:36–45.

6. Texas Instruments. 2006. MPS430x1xx Family User's Guide, SLAU049F. http://www. ti.com/litv/pdf/slau049f.

7. Texas Instruments. 2013. MPS430F5xx and MPS430x6xx Family User's Guide, SLAU208M. http://www.ti.com/lit/pdf/slau208.

8. Wilson, A. R., and P. S. Vincent. 2010. Networked low power sensing: Network interface and main operating system. *IEEE Sensors* 10:1495–1507.

9. NIST. 2011. Introduction to IEEE P1451. http://www.nist.gov/el/isd/ieee/1451intro. cfm.

10. Vincent, P. S., C. P. Gardiner, A. R. Wilson, D. Ellery, and T. Armstrong. 2008. Installation of a sensor network on an RAN Armidale class patrol boat. *Materials Forum* 33:307–16.

11. Gardiner, C. P., P. S. Vincent, A. R. Wilson, D. Ellery, and T. Armstrong. 2008. A trial sensor network for the Armidale Class patrol boat. Paper presented at Pacific 2008 International Maritime Conference, Sydney, Australia.

12. Bartholomeusz, R., B. Crawford, C. Davis, S. Galea, B. Hinton, G. McAdam, I. Powlesland, et al. 2004. DSTO aircraft structural prognostic health monitoring program for corrosion prevention and control. Paper presented at Corrosion & Prevention Conference. Australasian Corrosion Association Inc., Perth.

13. Trueman, T., P. Trathen, K. Begbie, L. Davison, and B. Hinton. 2007. The development of a corrosion prognostic health management system for Australian Defence Force aircraft. *Advanced Materials Research* 38:182–200.

14. Wilson, A. R., and R. F. Muscat. 2011. Novel thin wire paint and sealant degradation sensor. *Sensors and Actuators: A Physical*, doi:10.1016/j.sna.2010.10.012.

11 Maritime Data Management and Analytics: A Survey of Solutions Based on Automatic Identification System

Baljeet Malhotra, Hoyoung Jeung, Thomas Kister, Stéphane Bressan, and Kian-Lee Tan

CONTENTS

11.1 INTRODUCTION

The Automatic Identification System (AIS) is an important class of wireless technology that is being used for maritime navigation. Essentially, ships deploy the AIS to broadcast their navigational information captured through sensors such as GPS to other ships in their vicinity. That helps in enhancing the situational awareness for the primary purpose of safe navigation. The AIS operates in the very-high-frequency (VHF) maritime mobile band using Time Division Multiple Access technology [1]. A typical AIS setup consists of a transponder, a VHF antenna, and a GPS unit to capture the ship's own location. A collection of AIS transponders (connected to various onboard sensors in ships) creates a mobile wireless sensor network (WSN) that broadcasts data proactively, i.e., without the need of querying the network.

The AIS not only forms a communication network for exchanging information on navigation safety and security of ships, but it has also become a rich source of data on ships' identification, trajectories, navigation status, and other information that can be exploited to address some of the challenging problems in the maritime domain. To that end, many researchers have extensively used maritime data specifically generated by the AIS for trajectory data mining [2–9]. The AIS data have also been used to study a spectrum of multidisciplinary problems such as maritime emission control [10, 11], anomaly detection, risk assessments [12–14], complex network analysis [15], and others [16–18].

In this chapter, we offer an in-depth survey of the research proposals that have exploited maritime data specifically generated by the AIS to propose applications and data-analytic solutions for various problems. The main goal of this chapter is to compile and critique a comprehensive list of such research works, which can serve as an important basis in building new applications as well as offering a useful source of reference for various stakeholders, including port authorities, shipping companies, researchers, maritime practitioners, and law-enforcement agencies. This chapter summarizes the main ideas and techniques of the existing proposals while discussing the advantages and disadvantages of each proposal. In particular, we focus on the maritime data generated by the AIS and their applications in data analytics, a topic that has not been addressed in prior studies.

The rest of the chapter is organized as follows: In Section 11.2, we present an overview of the AIS and its main functions, followed by a brief description of an AIS station that we have set up at SAP Research in Singapore. In Section 11.3, we synthesize the related work in detail and present our views on the proposed applications and data-analytics solutions. Section 11.4 summarizes the chapter while discussing some important future research directions.

11.2 AUTOMATIC IDENTIFICATION SYSTEM

The International Maritime Organization (IMO) requires a majority of cargo and passenger ships to use the AIS [1] for the primary purpose of safe navigation. The AIS uses a VHF radio protocol. It allows ships and stations to broadcast messages in a specific binary-encoded form that contains a rich set of information, such as ships' Maritime Mobile Service Identity (MMSI) number as well as their position, navigational status,

TABLE 11.1

An NMEA 0183 Message and Some Decoded Data of the Embedded AIS Message

Type	Messages
Encoded	!AIVDM,1,1,,B,18Jjv:3PAg7K7R80dcC53l7j00T;,0*08
Decoded	MMSI#564969000; MsgType#1; Latitude#1.21954;
	Longitude#103.79062; SOG#11.1; COG#129.5; Heading#131

speed over ground (SOG), course over ground (COG), rate of turn (ROT), true heading (TH), and others, all for the primary purpose of navigation safety and traffic control.

When communicating with other marine electronic devices on board the ships, the AIS messages are encoded by using the NMEA 0183 standard [19]. A typical NMEA message (encoded) with its comma-separated fields is shown in Table 11.1. The sixth (and the longest) field is the actual AIS message that contains the ship information. In this encoding scheme, each ASCII character represents 6 bits of the original AIS message. The AIS messages are encoded primarily due to the limited bandwidth and also to limit the exploitation of the information (such as the position of the ships) by adversaries, which could seriously undermine the navigational safety and security of ships and their crew members.

11.2.1 AIS Communication Network

A typical scenario of a VHF-based communication network used by ships, VTS centers, and lighthouses for exchanging navigation, weather, and other related data is illustrated in Figure 11.1. Ships broadcast the AIS messages periodically using a common VHF channel. Position and other relevant data such as ships' SOG, COG, and ROT (captured through various sensing devices that are typically aboard ships) are automatically fed into the AIS messages. Lighthouses and VTS centers may also use the AIS messages to broadcast additional information to help ships during navigation. In particular, VTS centers can piggyback important information (on the AIS messages) such as local weather reports, birthing announcements, differential GPS (DGPS) corrections, and so on. A collection of AISs (equipped within ships, VTS centers, and lighthouses) that are periodically broadcasting messages can create a special scenario of a WSN in which some sensors (aboard ships) could be mobile and some stationary (e.g., sensors within a VTS), as illustrated in Figure 11.1. An important difference, though, as compared to a *regular* WSN is that the sensors (in an AIS-based WSN) are usually connected to a reliable source of power, and hence they are not constrained in terms of energy.

11.2.1.1 Frames and Slots

The AIS can broadcast 27 different types of encoded messages during various navigational situations. Refer to the literature [1] for more details on these messages. The AIS messages are broadcast using two VHF channels, i.e., 161.975 MHz and 162.025 MHz that are generally known as AIS1 and AIS2, respectively. The transponders

Lighthouses   VTS Control Centres   Additional Data Sources

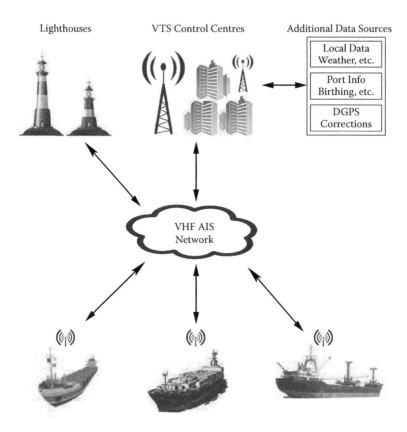

FIGURE 11.1 Scenario of a mobile communication network based on collection of AIS data.

use only one channel at any given time to transmit the messages. A message is first encoded using a non-return-to-zero inverted (NRZI) scheme and then transmitted through Gaussian-filtered minimum-shift-keying (GMSK) modulation at a rate of 9600 bps.

The AIS system divides the *broadcasting time* into frames in which every frame is equal to 1 minute. The clock is synchronized against the Coordinated Universal Time (UTC). Typically, a new frame starts at a new minute of the UTC *clock* (with the reference of a GPS signal). One frame is further divided into 2,250 slots. This means that at a rate of 9,600 bps, one slot is 256 bits long. Figure 11.2 summarizes this information and provides the details about the structure of an AIS message sent in a slot. Note that the AIS messages are encapsulated for radio transmissions and, as such, only 168 bits of data are sent in one slot.

11.2.1.2 Slots Allocation

The mechanism for slots allocation is nontrivial, requiring a detailed discussion that is beyond the scope of this chapter. For the sake of completeness, we summarize the

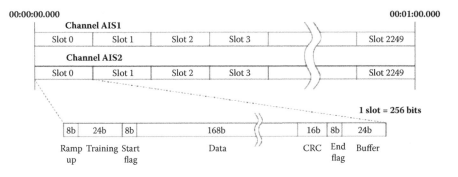

FIGURE 11.2 AIS frame and slots.

main idea here. A transponder allocates slots for itself and informs the others (in its neighborhood) through its AIS messages (typically the last part of the message is reserved for slots allocation). The other transponders can then act accordingly, knowing in which slot each transponder is going to transmit. Depending on the type of the messages and their importance, as well as the type of the transponders, different channel-access schemes are used:

- ITDMA, Incremental Time-Division Multiple Access
- RATDMA, Random Access Time-Division Multiple Access
- FATDMA, Fixed Access Time-Division Multiple Access
- SOTDMA, Self-Organizing Time-Division Multiple Access
- CSTDMA, Carrier-Sense Time-Division Multiple Access

Depending on the status of the ship (category, speed, and rate of turn), position report messages are sent from every 2 seconds to once every 3 minutes. Other messages are sent every 6 minutes. Moreover, the base stations (e.g., VTS) can query ships about their position at any time. Once again, the full reference is available in the literature [1].

11.2.1.3 Data Rates

Usually, if the AIS message is short enough, it occupies only one slot. Otherwise, it is transmitted using multiple slots. If the data in a slot is shorter than 168 bits, then it is padded with extra (unused) bits. Because each channel can accommodate 2250 slots per minute, the maximum number of messages that can be transmitted is 4500 per minute. However, since the AIS messages could be of different lengths depending on their types, one can take the size of the data into account rather than the number of messages. In a typical AIS communication network, we can expect to have a maximum data rate of $R_{AIS_{max}} = 9,600 \times 2 \times (168/256) = 12,600$ bps. That is about 129.78 MB/day or 3.8 GB/month or 46.26 GB/year.

It is interesting to note that the AIS communication network provides a mechanism to fall back on regional reserved frequencies in case of congestion on the AIS1 and AIS2 channels. However, the current implementation status of this mechanism is unknown to us. We suspect that it has not been implemented yet in the zone that we are currently covering. It is intuitive that the special mechanism to overcome the congestion may require special hardware/software setup, which our current receiver can not accommodate. Also, like any other broadcasting radio system, an AIS communication network suffers from the *hidden terminal* problem as well. For instance, if a receiver is positioned between two transmitters that are not within each other's transmission range, then their messages (if transmitted during the same time slot) may get corrupted without any chance of recovery. Furthermore, the transmission power of the equipment is not necessarily fixed, and consequently it is possible for some ships to go undetected because of their low transmission power. On the other hand, obstacles such as a building on an island and interferences generated by electronic equipment can potentially hamper the communication of an AIS network. We do not intend to address these issues in this chapter, but they will be investigated in our future research.

11.2.2　AIS DATA MANAGEMENT

Stream Data Management Systems (SDMS) [20] have recently gained importance due to their applications in telecommunications, health care, transportation, and so on. The AIS can also be seen as a source of data streams consisting of special decoded messages, as shown in Table 11.1. The AIS produce a continuous stream of such messages containing a rich amount of information about ships. The AIS messages can be received and decoded by other ships and by Vehicular Traffic Service (VTS) stations in the vicinity (of approximately 20–40 nautical miles). Indeed, the AIS messages can be received and decoded by anyone in the neighborhood (of transponders) who is equipped with an appropriate receiver and decoding hardware and software.

The AIS data can be collected from an AIS communication network while using a multilayer system typically consisting of a Complex Event (Stream) Processing Engine (CEP) [21, 22] and a database system for processing, storing, and analysis of the AIS data. At SAP Research, we have set up an AIS station to collect data from ships arriving at the port of Singapore. The overall infrastructure of the setup is shown in Figure 11.3. The captured AIS data is being processed and analyzed using EventInsight©, which is a CEP from SAP. CEPs usually do not store data permanently; however, they allow access to traditional databases such as Oracle© and SQL Server© for data storage and processing purposes. As shown in Figure 11.3, we interfaced EventInsight with HANA©, which is SAP's in-memory database appliance.

Through this setup, we not only intend to demonstrate the management, processing, and analysis of maritime data generated by ships, but also to allow the development of innovative solutions for various stakeholders. This partially completed infrastructure is currently facilitating the research and development of new applications and analytics solutions for various stakeholders and partners of SAP. A detailed discussion on the infrastructure, including hardware and software components, is not within the scope of this chapter. Rather, we focus on solutions that have been proposed for various problems using the AIS data sets. To that end, in the following section we discuss related work that has exploited the AIS data sets to propose various analytics solutions.

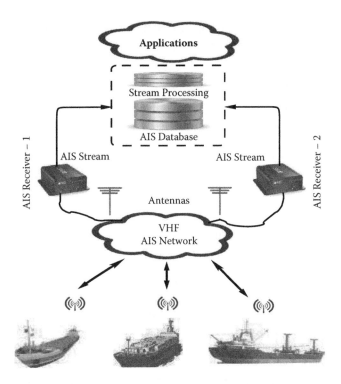

FIGURE 11.3   Scenario of AIS data management and related applications.

11.3   AIS DATA MANAGEMENT AND ANALYSIS

Many existing proposals have exploited the AIS data sets for various applications, such as maritime traffic management [3–5], anomaly detection based on moving patterns of ships [12–14], impact of marine vessel emissions [10, 11, 23], stream data management [11, 16], and others [24–28]. Many of these applications require analytical techniques for mining [29–33], clustering [2, 6–9, 34, 35], joining [36–41], outlier detection [42], and modeling [3, 43, 44] to analyze the AIS data sets in an effort to find patterns and trends.

Perez, Chang, and Billings [11] discussed the challenges of data management, analysis, and the problems of missing data in the AIS data sets while proposing potential methods for addressing these limitations. Yet another study [16] discusses the management of the AIS data streams from the perspective of privacy and access control. Unlike the proposals in these two works [11, 16], we do not intend to propose new solutions to address these problems or the limitations of the existing solutions. Rather, in this chapter, we summarize the proposed solutions while classifying them into various categories based on the analytical solutions used as well as the context of the applications in which those analytical solutions have been proposed. Toward that end, we use four main categories to synthesize the related work: (1) trajectory data mining, (2) maritime traffic management and anomaly detection, (3) marine environmental health monitoring, and (4) privacy and access control. We start our discussion with the trajectory data-mining proposals.

11.3.1 TRAJECTORY DATA MINING

Trajectory data mining is an emerging and rapidly developing topic in the areas of data mining that aims at discovering patterns of trajectories based on their proximity in either a spatial or a spatiotemporal sense. As ships keep moving and continuously generate trajectory data reported by GPS integrated with the AIS, mining their trajectories plays an important role in maritime data management [2, 30, 42, 43 45, 46]. For instance, at a commercial port where hundreds of vessels may enter or leave the port or wait to do so, collision avoidance is of utmost importance [24].

Data-mining methods can be employed to mine trajectory data collected from the vessels to discover mobility, traffic, and congestion patterns, which can then be used for situational awareness [43]. Based on the movement patterns, trajectories (and the vessels spanning them) can be clustered into groups to access the interactions between them and their collision risks [2, 44]. Furthermore, models can be built to engineer monitoring systems based on the discovered patterns, such as the one proposed by Piciarelli and Foresti [47]. Such models can then be used to detect anomalies (e.g., the trajectory of a particular ship that is not adhering to the guidelines) in real time to warn the authorities in advance.

There are a rich variety of methods that enable the discovery of important knowledge from vessel-driven trajectory data. The following subsections present some of the key data mining classes that play potentially important roles in maritime data analysis.

11.3.1.1 Periodic Pattern Mining

Periodic pattern mining of trajectory data concerns the discovery of periodic object behavior [32, 33], i.e., objects that follow the same routes (approximately) over regular time intervals. These periodic patterns provide an insight into, and concise explanation of, periodic behaviors (e.g., daily, weekly, monthly, and yearly) across long movement histories. Periodic patterns are also useful for compressing movement data [29], since they summarize movement trajectories into a compact format. In addition, periodic patterns can serve as a basis for predicting future movements [31]. Moreover, if an object fails to follow an established, regular behavior, this could be a signal of an abnormal environmental change or an accident.

When considering object movement, it is typically unreasonable to expect an object to repeat its behavior exactly during each time period considered. This implies that the patterns to be identified should not be rigid; rather, that object's behavior should be allowed to differ slightly from one period to the next while still resulting in an overall pattern. Next, behaviors that make up patterns may also be shifted in time (e.g., due to traffic delays). The approximate nature of patterns in the spatiotemporal domain makes the mining tasks even more difficult. For example, it is hard to discover a time-relaxed driving pattern across different objects that move with slightly different departure times as well as speeds, as the discovery process needs to take into account adjacent data points in time. In addition, the periods that yield patterns may be unknown. Further, there may be multiple periods (e.g., day and week) that yield different patterns in the same data. As a result, periodic pattern mining of trajectory data takes into account a wide variety of modeling approaches as well as efficient discovery algorithms.

11.3.1.2 Trajectory Join

Some trajectory patterns are defined and computed by means of database join queries. Given two data sets P_1 and P_2, spatiotemporal joins find pairs of elements from the two sets that satisfy a given predicate with both spatial and temporal attributes [36, 38]. The study of vessel trajectories with the objective of finding incidents may be accomplished using joins. Since joins may involve the comparison of all trajectories in data set P_1 with all trajectories in data set P_2, which is computationally expensive, a common approach for join processing involves the use of indexing techniques to avoid unnecessary distance computations.

The *close-pair join* [37] reports all object pairs (o_1, o_2) from $P_1 \times P_2$ with distance $D_\tau(o_1, o_2) \le e$ within a time interval τ, where e is a user-specified distance. Plane-sweep techniques [40, 41] have been proposed for evaluating spatiotemporal joins. Zhou et al. [41] use join predicates that define a rectangular region in time and space. An index structure (MTSB-tree) is introduced to enable efficient retrieval of the pairs of trajectories that satisfies the join predicates. Instead of using an index, Arumugam and Jermaine [40] utilize MBR approximations of trajectory segments to reduce the computation of query processing.

Like the close-pair join, the *trajectory join* [39] aims at retrieving all pairs of similar trajectories in two data sets. Bakalov et al. [39, 48] represent trajectories as sequences of symbols and apply sliding-window techniques to measure the symbolic distance between possible pairs.

11.3.1.3 Clustering of Trajectories

Clustering of trajectories is perhaps one of the most fundamental operations used in various types of trajectory-pattern mining, since the discovery of trajectory patterns typically involves the process of grouping similar positions, trajectories, and objects. The key concepts in state-of-the-art trajectory clustering techniques are as follows:

- Kalnis et al. introduced the concept of a *moving cluster* [34], which is defined as a set of objects that move close to each other for a time duration. It is a sequence of spatial clusters appearing during consecutive time points, such that the portion of common objects in any two consecutive clusters is not below a given threshold θ, i.e., $\frac{|c_t \cap c_{t+1}|}{|c_t \cup c_{t+1}|} \ge \theta$, where c_t denotes a cluster at time t.
- Lee, Han, and Whang [9] present another approach for trajectory clustering, called TRACLUS, for the purpose of grouping trajectory segments. Specifically, they proposed a partition-and-group framework *TRACLUS* that clusters trajectories by dividing raw trajectories into trajectory segments and then groups the segments.
- The concept of *convoy* [7, 8] employs the notion of density connection [35] to enable the formulation of arbitrary shapes of groups. More specifically, given a set of trajectories O, an integer m, a distance value e, and a lifetime

k, a *convoy* is defined as a group that has at least m objects that are density-connected with respect to distance e and cardinality m during k consecutive time points. Each convoy is associated with a time interval during which the objects in the group traveled together.

- Li et al. [6] introduced a new trajectory pattern type, *swarm*, that extends the concept of convoy by relaxing the consecutive-time constraint. Specifically, the definition of *swarm* replaces the parameter k of *convoy* with k_{min}, such that k_{min} denotes a minimum of time duration to form a moving object cluster, regardless of the consecutiveness in time. This allows an individual moving object to temporarily leave its group as long as it is close to other group members during most of the time.

- Li et al. [2] addressed the problem of incremental clustering of trajectories and their visualization. Because trajectories are sequential in nature, i.e., the trajectory data are often received incrementally, an incremental clustering framework was proposed, which contains two parts: (1) online micro-cluster maintenance and (2) off-line macrocluster creation. In their online approach, trajectories are first simplified into a set of directed line segments to find clusters of trajectory subparts. Microclusters are used to store compact summaries of similar trajectory line segments, which take much less space than raw trajectories. When new data are added, microclusters are updated incrementally to reflect the changes. In the offline part, when a user requests to see current clustering results, macroclustering is performed on the set of microclusters rather than on all trajectories over the whole time span. Since the number of microclusters is smaller than that of original trajectories, macroclusters are generated efficiently to show clustering results on trajectories.

11.3.1.4 Semantic Interpretation and Feature Selection

A significant challenge posed by the trajectory data is the semantic interpretation of the background geographical areas spanned by the corresponding trajectories. Some existing solutions [30] rely on mining techniques and/or generic databases, e.g., a GIS map, to find regions of interest. However, as noted by Alvares et al. [49], such solutions are ad hoc and may not necessarily produce desired results. For instance, finding a certain percentage of the trajectories that pass through a particular region of interest may indicate the popularity of that region, but without providing any interesting insights that reveal the characteristics of that region. This is partly because the background geographical information is often-times not part of the trajectory data. Nonetheless, the background geographical information is fundamental to the mining and analysis of the trajectory data for a better interpretation of the results. To that end, some solutions have been proposed that take semantic interpretation (of the background geographical information) into account. Specifically, Alvares et al. [43] proposed a reverse-engineering approach for understanding the patterns in the trajectory data while accounting for the semantics and the background geographic information. Their basic approach is to model the trajectory patterns in a database schema that is drawn from relevant geographical information. However, none of the solutions [43, 49] evaluate the impact of *attaching* semantics on the data mining of these semantically rich trajectories.

Andrienko, Andrienko, and Wrobel [45] argued in favor of visual analytics tools that could be useful for managing large amounts of movement data, such as that generated by the AIS. Their primary objective was to embed the necessary semantic information within movement data, which could then facilitate in designing new data-mining algorithms for efficient analysis of movement data. The proposed techniques have been demonstrated using a real AIS data set.

Yet another challenge posed by trajectories is the volume of the data in space and time. In a typical situation, trajectories may span vast geographical regions or time periods or both to significantly increase the volume of the data. Mining such high-dimensional data to discover useful patterns is not a trivial problem. In the context of maritime vessel-to-vessel interactions, it could hamper the extraction of meaningful patterns. A typical approach taken in this situation is to *scale down* the trajectories by means of selecting only the prominent features in order to reduce the dimensionality [50].

De Vries, van Hage, and van Someren [46] presented a similarity measure that combines the low-level trajectory information with geographical domain knowledge to compare vessel trajectories. The proposed similarity measure was largely based on the alignment techniques. Using these techniques, the discovery of behavior patterns was demonstrated from trajectory data that are dependent both on the low-level trajectories as well as the domain knowledge. The proposed similarity measure was eventually used to predict the type of vessels.

11.3.2 Maritime Traffic Management and Anomaly Detection

Discovering knowledge from the AIS database for applications in maritime traffic management has been the focus of many recent studies [3–5]. The AIS enables the Vessel Traffic Service (VTS) and port authorities not only to offer commonly known functions such as identification, tracking, and monitoring of vessels, but also to provide rich real-time information that is useful for marine traffic investigation, anomaly and threat detection, statistical analysis, and theoretical research.

Goerlandt and Kujala [3] used the AIS data to build analytical models for predicting ship collisions. In particular, they proposed a method to assess the probability of vessels colliding with each other at a given location and the time when collisions are most likely to occur. The proposed method is based on time-domain microsimulation of vessel traffic that is observed through the AIS. The Monte Carlo simulation technique was applied to predict the relevant factors of the collision events.

Tsou [4] applied data-mining techniques used for business intelligence discovery (e.g., as in customer relation management for business marketing) to the analysis of the AIS data. This was done by mapping the marine traffic problem as a business-marketing problem and by integrating technologies such as Geographic Information Systems (GIS), database management systems, data warehousing, and data mining. The eventual goal was to provide the marine traffic managers with a useful strategic planning resource that could facilitate the discovery of hidden and valuable information from a huge amount of data.

Tang and Shao [5] presented a survey on marine traffic. The purpose was to gather and analyze basic data by using all effective means to know the traffic status, characteristics, and general rules at both macroscopic and microscopic levels. Based on the eigenvalues obtained through the AIS data, the marine traffic was analyzed by means of clustering and statistics. Their proposed framework further supports the use of data mining in the forecast of marine traffic flow and in the development and programming of marine traffic engineering.

Ristic et al. [12] took a two-phase approach of motion anomaly detection and motion prediction to address the problem of maritime anomalies. In the first part, a kernel density estimation (KDE) was used on the historical AIS data to separate the normal motion behavior of ships from the abnormal ones. The primary objective was to determine a threshold value that could distinguish the incoming *unknown* motions of ships into normal and abnormal motions, which was the second aspect of their work. The motion of vessels was predicted using the Gaussian sum-tracking filter. The proposed techniques were demonstrated using a real AIS data set collected in Adelaide and Sydney (Australia).

Laxhammar, Falkman, and Sviestins [13] evaluated two previously proposed methods for statistical anomaly detection in sea traffic, namely the Gaussian Mixture Model (GMM) and the adaptive Kernel Density Estimator (KDE). It was claimed that, due to the combinatorial nature of the problem, the computational complexity for finding the adaptive window widths and anomaly detection are, respectively, quadratic and linear in the size of a training data set. To solve this problem, a discretization of the geographical area (spanned by the corresponding trajectories) was proposed by dividing it into multiple cells.

Jakob et al. [14] explored agent-based techniques to assess the threat of contemporary maritime piracy to international transport. The main idea of the approach was to build a computational model on maritime activity by using a range of real-world data sources. The model applies statistical machine-learning techniques to extract patterns of vessel movement from trajectory data; the models were subsequently used for categorizing vessels and detecting suspicious activities. Another module, which employed game theory–based strategic reasoning, was used to plan risk-minimizing routes for vessels transiting through known pirate waters. A simulation platform was eventually used for advanced analysis, reasoning, and planning capabilities.

Outlier detection is another technique that has been exploited to find anomalies in trajectory data. Lee, Han, and Li [42] proposed a partition-and-detect framework for detecting outliers. The basic idea of the proposal was to partition a trajectory into a set of line segments from which the *outlying line segments* were classified as outliers. A hybrid of the distance-based and density-based approaches was used to classify outlying line segments.

Laxhammer, Falkman, and Sviestins [13] defined two performance metrics, *normalcy modeling performance* and *anomaly detection performance*, for evaluating the Gaussian Mixture Model (GMM) and the adaptive Kernel Density Estimator (KDE) methods for trajectory analysis. In terms of normalcy modeling performance, a method was considered to be superior if it was able to better estimate the true

Probability Density Function (*PDF*) of an unknown trajectory sample. In terms of anomaly detection performance, a method was considered to be superior if it was able to classify an anomalous trajectory while using a minimum number of trajectory observations (i.e., data points that construct a trajectory segment). Using these criteria, it was concluded that KDE is a more appropriate technique to accurately capture the features in normal data to detect anomalies, e.g., in this particular study, the spatial distribution of the vessel location along sea lanes. The results from anomaly detection, however, show no significant differences between the two techniques.

Zhu [51] applied an electronic chart system, database management, and data warehouse and data mining technologies to facilitate the discovery of hidden and valuable information in a huge amount of the AIS data. Association Rules [19] were applied to mine the ship trajectories in an effort to find patterns. It was observed that the mined knowledge was indeed useful for their maritime safety administration.

Overall, most of the approaches [3–5, 12–14, 42, 51] discussed in this section apply various statistical techniques to discover the marine traffic patterns. In particular, two techniques—the Gaussian Mixture Model (GMM) and the adaptive Kernel Density Estimator (KDE)—have been found to be useful for distinguishing the abnormality in the marine traffic patterns. Some of the proposed solutions [12, 46] extended their features to perform the classification tasks to predict the moving behavior and the class/type of ships. Due to the combinatorial nature of the problem, feature selection for the purpose of classification poses several challenges. However, there seems to be no consensus among the existing solutions on any particular feature-selection scheme that could be useful for the purpose of classification.

11.3.3 Marine Environmental Health Monitoring

Information from databases on ships combined with the AIS-based data sets have been exploited in existing proposals [10, 11, 23] to estimate ships' pollution emissions for monitoring the health of marine environment. The basic idea of these proposals is to estimate the energy (fuel) consumed by ships while they are sailing, maneuvering, mooring, or performing any other activities that can be observed through the AIS stations. Intuitively, if the details of ships' engines (and those of other mechanical/ electrical parts) are available, then one can estimate the total energy consumed by ships under various conditions. Regardless of the challenges in data management and analysis, and the problems of missing/incorrect data [11], the AIS provides the opportunity for highly refined vessel movement and improved estimation of emissions [10].

Pitana, Kobayashi, and Wakabayashi [10] conducted research in which the basic aim was to evaluate the marine traffic contributions to the nitrogen oxides (NO_x), sulfur oxides (SO_x), particulate matter (PM), carbon monoxide (CO), and carbon dioxide (CO_2) levels in the Madura Strait area, which is one of the busiest marine traffic regions in Indonesia. It was observed that many ships arrive, depart, and travel through that area, all of which influence the air quality of the port environment. At the time of the study, the Indonesian government had not yet ratified any policy regarding the prevention of air pollution from ships. It was suspected that the absence of any restricting regulations could have an influence on the level of air pollution. To investigate these issues, a decision-making tool was used to evaluate the effects of ship emissions on port/marine health. To that end, three types of data were extracted from the AIS data sets—speed of ship, initial position of ship, and ship type apart from MMSI number—which were then used for analysis.

In a similar but separate study, Perez, Chang, and Billings [11] at Eastern Research Group, Inc. (ERG) used the AIS data sets to create a state-of-the-art inventory of 2007 commercial marine vessel emissions in waters of Texas. To that end, a geographic information system (GIS) was used to map and analyze both individual ship movements and general traffic patterns on inland waterways and within 9 miles off the Texas coastline. ERG then linked the vessel-tracking data to individual vessel characteristics from Lloyd's Register of Ships, the American Bureau of Shipping, and Bureau Veritas to match vessels to fuel and engine data. These results were then applied to the latest emission factors to quantify criteria and hazardous air pollutant emissions from the vessels observed through the AIS station.

Based on the Ship Traffic Emissions Assessment model at the Finnish Meteorological Institute, Johansson [23] presented an extended method for the evaluation of the exhaust emissions of marine traffic. The proposed model uses the AIS data (mainly for extracting ships' routes and their speed) apart from the engine load, fuel sulfur content, multiengine setups, abatement methods, and waves. Compared to the existing solutions, the proposed model can compute the mass-based emissions of particulate matter (PM) and carbon monoxide (CO). The author also built modules to process the AIS data before it is used for modeling.

In an interesting report from the Federal Environment Agency (Umweltbundesamt) of Germany, Bäuerle et al. [52] presented the state of the art on legalities and challenges of using AIS data for the purpose of emission estimations. It was argued that the record of distance sailed could be falsified easily when no secondary control is used to verify the reporting. Therefore, an automated reporting system such as the AIS serves as a secondary backup of the fuel consumption reporting (the primary being the bunker fuel delivery notes or log-book entries). The fuel consumption and the corresponding emissions of carbon dioxide can then be calculated, which are mainly related to the vessel size, its service speed, and the hull design. A bottom-up modeling technique is used to estimate the emissions, which is based on the vessel's engine power (main and auxiliary), load-factor assumptions, resulting speed calculations, days at sea and days in port, fuel consumption allocations based on size and age of the vessel, and the use of emission factors for the used fuels.

11.3.4 MODELING AND SIMULATION

Modeling the dynamics of ships is a challenging problem. Generally, a close proximity between two or more ships leads to hydrodynamic interactions between them [28]. Such interactions are common, especially in dense maritime traffic, where meeting and overtaking maneuvers are unavoidable. These interactions may also occur during escorting or maneuvering of vessels for berthing assistance at sea ports. According to a recent study described in the *Review of Maritime Transport* [53], global seaborne trade will increase by 44% in 2020, and it will double by 2035. An increase in global sea trade will naturally increase the number of vessels (and their transportation capacity), resulting in an increase of marine traffic, which in turn is bound to affect the marine traffic patterns and the port operations. To that end, many recent studies have been dedicated to studying the dynamics of ships [17, 25–27]. Next, we discuss some proposals in which the analytics based on the AIS data play an important role for modeling the dynamics of ships.

- Usually, simulation models are used to evaluate the maneuvering and course-keeping of ships. That requires a deep understanding of the hydrodynamic forces, which may affect ships' maneuverability in waves [25–28]. Building better simulation models requires a systematic approach to understanding the relationship between ships' trajectories. To that end, data-mining solutions (which we have already reviewed in the previous sections) are required for extracting patterns from trajectories (which could be extracted from the AIS data sets). Intuitively, extracting patterns from ships' trajectories can further enrich the existing models that predict the behavior of ships in a dynamic environment. Several modeling techniques, such as verbal models and mechanical analogies, fuzzy subsets, statistical models, difference and differential equations, and stochastic models exist [28] that can be used for modeling the dynamics of ships. In this context, the historical AIS data is useful to validate the models that predict certain aspects of ships' dynamics.
- *Wake-fields* are formed when vessels pass through the medium, and these enforce a limit on vessels' proximity to other vessels, which must be maintained for safe navigation [17]. It essentially means that every ship has a *safety domain*, which is basically an area around a ship that must not be intruded by any other ships during navigation. Unfortunately, interactions between ships are unavoidable in many situations, e.g., during port calls or in narrow water channels. In this situation, the historical AIS data could be useful in detecting the safety domain violations. Analytics on domain violations, e.g., to find: (1) what types of vessels cause more violations than others, (2) in which geographical regions (that are bottlenecks) violations are more frequent, and (3) during which particular seasons violations occur more frequently, etc., could serve as key performance indicators on marine traffic management.

In an interesting study, Harati-Mokhtari et al. [18] investigated the human errors that occur with the AIS and their impact on navigation safety. Their study found that, in several cases, the AIS data were not reliable. It was also found that many of the input errors in the AIS related to ships' navigational status were due to memory slips or omissions to execute an action. The conclusion of the study was that proper supervision, surveillance of accuracy, and enforcement of quality of the AIS data by competent maritime authorities could enhance its efficacy in all navigation operations.

11.3.5 PRIVACY AND ACCESS CONTROL

With a network of AIS transponders making data available on the Internet, one can combine multiple AIS data streams to offer various services based on a global view of maritime ships for both operational and analytical purposes. For example, consider the scenario depicted in Figure 11.4, in which Singapore is shown to be under the surveillance of multiple AIS receivers installed at various locations along the island to maximize the coverage. Such AIS-based services could be useful to port authorities, shipping and insurance companies, cargo owners, and other stakeholders. Consider the following three examples: (1) A shipping company may track its vessels around the world and analyze their routes; (2) a ship captain (more accurately, a second officer, responsible for navigation watch) can be informed of the positions and movements of ships that are relevant to safe navigation; (3) a port authority can complement its traffic control by monitoring ships calling at the port and navigating the port waters. Individuals or organizations deploying the appropriate AIS receiver and decoding equipment can locally monitor ships.

Unfortunately, the free and uncontrolled distribution of AIS data on the Internet can be exploited to seriously undermine navigation safety and security. For instance, it can be used by pirates to plan and launch long-range attacks. It can also violate the privacy of the various stakeholders. For instance, it can reveal information about the cargo and the cargo owners. This problem has been highlighted by IMO's Maritime Safety Committee, which declared that

> the publication on the world-wide web or elsewhere of AIS data transmitted by ships could be detrimental to the safety and security of ships and port facilities and was undermining the efforts of the Organization and its Member States to enhance the safety of navigation and security in the international maritime transport sector. [54]

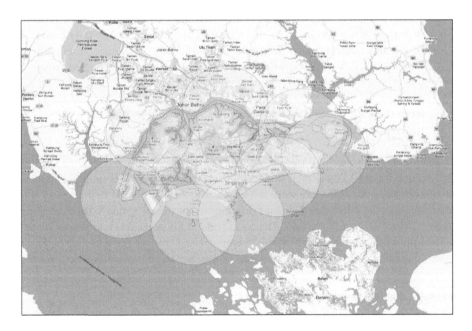

FIGURE 11.4   Scenario of multiple AIS receivers covering the coastal waters of Singapore.

In an important study, Bäuerle et al. [52] also argued that the AIS data streams essentially contain private and/or sensitive information that concern not only the privacy of various stakeholders, but also raise several legal issues regarding their usage. Following the United States' control policy on scientific and technical information [55], the AIS data streams could be described as Sensitive But Unclassified (SBU). In this context, the access and accessibility to SBU data streams such as the one generated by AIS needs to be controlled to ensure their legitimate usage in a secure and private manner.

Though the AIS data streams have been exploited for various purposes (e.g., in proposals [47, 56, 57] as discussed in the previous sections), not much attention has been paid to securing the AIS data streams with access-control policies. Fortunately, in the existing literature on stream data management systems (SDMS), a lot of attention has been paid to access control of data streams [58, 59]. A scenario of access control over the AIS data streams is shown in Figure 11.5. Raw encoded messages are drawn from the AIS network into a SDMS, which is then used to manage the AIS data (queries), and also to control the AIS data streams to provide access to authorized users only. Malhotra et al. [16] demonstrated various scenarios of access control over the real AIS data streams.

Because of their usefulness in analytics solutions, the AIS data sets are available in the commercial market for purchase [60]. These data sets are mostly historical in nature. Access control can not only facilitate in securing the privacy of authorized

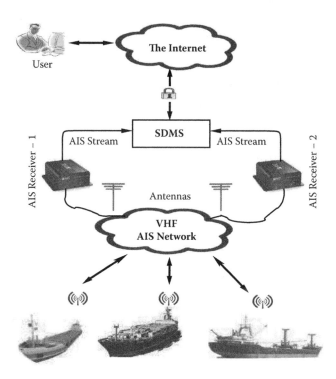

FIGURE 11.5   Scenario of access control over AIS data streams.

users, but it can also offer a framework for the monetization of live data streams. Organizations are constantly looking into leveraging their data streams to create new business opportunities [61]. In the context of live AIS data streams, location-based service providers could benefit from such monetization.

In summary, privacy and access control of AIS data streams are important areas of research whose significance has been recognized by several European and North American countries. Satellite-based AIS systems may further complicate the privacy and access-control issues [52]. Nonetheless, the monetization of live AIS data streams will continue to be an active area for research and commercialization.

11.4 CONCLUSIONS

The AIS is a special class of wireless communication system that is primarily used by ships to enhance navigation safety. In this chapter, we surveyed recent proposals on maritime applications and data analytics solutions that exploited maritime data specifically generated by the AIS. To synthesize the related work, we summarized the proposed solutions into four main categories based on the proposed analytical techniques as well as the context of the applications: (1) trajectory data mining, (2) maritime traffic management and anomaly detection, (3) marine environmental health monitoring, and (4) privacy and access control. Our synthesis of the existing solutions reveals that the trajectory data-mining area has matured significantly, which can become a building block for creating new applications, especially in the areas of data analytics and monetization of the AIS data streams.

ACKNOWLEDGMENTS

The authors would like to thank the editors for their valuable suggestions. This research was partially supported by SAP Research, Economic Development Board (EDB) of Singapore, and the A*Star SERC project, "Hippocratic Data Stream Cloud for Secure, Privacy-preserving Data Analytics Services," 102 158 0037, NUS Ref:R-702-000-005-305.

REFERENCES

1. International Telecommunications Union. 2007. Technical characteristics for an automatic identification system using time division multiple access in the VHF maritime mobile band. ITU-R M.1371-3.
2. Li, Z., J.-G. Lee, X. Li, and J. Han. 2010. Incremental clustering for trajectories. In *Proceedings of the 15th International Conference on Database Systems for Advanced Applications*, vol. 2, 32–46. Berlin: Springer-Verlag.
3. Goerlandt, F., and P. Kujala. 2011. Traffic simulation based ship collision probability modeling. *Reliability Engineering and System Safety* 96 (1): 91–107.
4. Tsou, M.-C. 2010. Discovering knowledge from AIS database for application in VTS. *Journal of Navigation* 63 (3): 449–69.
5. Tang, C., and Z. Shao. 2009. Data mining platform based on AIS data. In *International Conference on Transportation Engineering*, vol. 5, 449–69. Reston, VA: ASCE.

6. Li, Z., B. Ding, J. Han, and R. Kays. 2010. Swarm: Mining relaxed temporal moving object clusters. *Proceedings of the VLDB Endowment* 3 (1): 723–34.

7. Jeung, H., M. L. Yiu, X. Zhou, C. S. Jensen, and H. T. Shen. 2008. Discovery of convoys in trajectory databases. *Proceedings of the VLDB Endowment* 1 (1): 1068–80.

8. Jeung, H., H. T. Shen, and X. Zhou. 2008. Convoy queries in spatio-temporal databases. In *Proceedings of the IEEE 24th International Conference on Data Engineering (ICDE)*, 1457–59. Piscataway, NJ: IEEE Press.

9. Lee, J., J. Han, and K. Whang. 2007. Trajectory clustering: A partition-and-group framework. In *Proceedings of the ACM International Conference on Management of Data (SIGMOD)*, 593–604. New York: ACM Press.

10. Pitana, T., E. Kobayashi, and N. Wakabayashi. 2010. Estimation of exhaust emissions of marine traffic using Automatic Identification System data (case study: Madura Strait area, Indonesia). In *Proceedings of the IEEE OCEANS*, 1–6. Piscataway, NJ: IEEE Press.

11. Perez, H., R. Chang, and R. Billings. 2009. Automatic identification systems (AIS) data use in marine vessel emission estimation. Paper presented at 18th Annual International Emission Inventory Conference. http://www.epa.gov/ttn/chief/conference/ei18/session6/perez.pdf.

12. Ristic, B., B. LaScala, M. Morelande, and N. Gordon. 2008. Statistical analysis of motion patterns in AIS data: Anomaly detection and motion prediction. In *11th International Conference on Information Fusion*, 40–46. Piscataway, NJ: IEEE Press.

13. Laxhammar, R., G. Falkman, and E. Sviestins. 2009. Anomaly detection in sea traffic: A comparison of the Gaussian mixture model and the kernel density estimator. In *12th International Conference on Information Fusion*, 756–63. Piscataway, NJ: IEEE Press.

14. Jakob, M., O. Vanek, S. Urban, P. Benda, and M. Pechoucek. 2010. Employing agents to improve the security of international maritime transport. In *Proceedings of 6th Workshop on Agents in Traffic and Transportation*, 29–38. Oxford, England: Elsevier.

15. Kaluza, P., A. Kölzsch, M. T. Gastner, and B. Blasius. 2010. The complex network of global cargo ship movements. *Journal of the Royal Society Interface* 7:1093–1103.

16. Malhotra, B., W.-J. Tan, J. Cao, T. Kister, S. Bressan, and K.-L. Tan. 2011. Assist: Access controlled ship identification streams. In *Proceedings of the 19th ACM SIGSPATIAL International Conference on Advances in Geographic Information Systems (GIS)*, 485–88. New York: ACM Press.

17. Yasukawa, H. 2006. Simulation of wave-induced motions of a turning ship. *Journal of the Japan Society of Naval Architects and Ocean Engineers* 4:117–26.

18. Harati-Mokhtari, A., A. Wall, P. Brooks, and J. Wang. 2007. Automatic Identification System (AIS): Data reliability and human error implications. *Journal of Navigation* 60 (3): 373–89.

19. National Marine Electronics Association (NMEA). 2008. Publications and Standards: NMEA 0183. http://www.nmea.org/content/nmea_standards/nmea_standards.asp.

20. Muthukrishnan, S. 2003. Data streams: Algorithms and applications. In *Proceedings of the 14th ACM-SIAM Symposium on Discrete Algorithms*, 413. Philadelphia: Society for Industrial and Applied Mathematics.

21. Arasu, A., B. Babcock, S. Babu, M. Datar, K. Ito, I. Nishizawa, J. Rosenstein, and J. Widom. 2003. STREAM: The Stanford stream data manager. In *Proceedings of the 2003 ACM SIGMOD International Conference on Management of Data*, 665. New York: ACM Press.

22. Abadi, D. J., D. Carney, U. Çetintemel, M. Cherniack, C. Convey, S. Lee, M. Stonebraker, N. Tatbul, and S. B. Zdonik. 2003. Aurora: A new model and architecture for data stream management. *Proceedings of the VLDB Endowment* 12 (2): 120–39.

23. Johansson, L. 2011. Emission estimation of marine traffic using vessel characteristics and AIS data. MSc. thesis, Aalto University, Finland.

24. Statheros, T., G. Howells, and K. McDonald-Maier. 2008. Autonomous ship collision avoidance navigation concepts, technologies and techniques. *Journal of Navigation* 61:129–42.

25. Ankudinov, V. 1983. Simulation analysis of ship motion in waves. Paper presented at International Workshop on Ship and Platform Motions, Berkeley, CA.

26. Ottosson, P., and L. Bystrom. 1991. Simulation of the dynamics of a ship maneuvering in waves. *Trans. Society of Naval Architects and Marine Engineers* 99:281–98.

27. Fossen, T. I. 2002. *Marine control systems: Guidance navigation and control of ships rigs and underwater vehicles.* Trondheim, Norway: Marine Cybernetics.

28. Aris, R. 1994. *Mathematical modelling techniques.* Mineola, NY: Dover Publications.

29. Cao, H., N. Mamoulis, and D. W. Cheung. 2007. Discovery of periodic patterns in spatiotemporal sequences. *IEEE Trans. Knowledge and Data Engineering* 19:453–67.

30. Giannotti, F., M. Nanni, F. Pinelli, and D. Pedreschi. 2007. Trajectory pattern mining. In *Proceedings of the ACM International Conference on Knowledge Discovery and Data Mining (SIGKDD)*, 330–39. New York: ACM Press.

31. Jeung, H., Q. Liu, H. T. Shen, and X. Zhou. 2008. A hybrid prediction model for moving objects. In *Proceedings of the IEEE International Conference on Data Engineering (ICDE)*, 70–79. Piscataway, NJ: IEEE Press.

32. Li, Z., B. Ding, J. Han, R. Kays, and P. Nye. 2010. Mining periodic behaviors for moving objects. In *Proceedings of the ACM International Conference on Knowledge Discovery and Data Mining (SIGKDD)*, 1099–1108. New York: ACM Press.

33. Mamoulis, N., H. Cao, G. Kollios, M. Hadjieleftheriou, Y. Tao, and D. W. Cheung. 2004. Mining, indexing, and querying historical spatiotemporal data. In *Proceedings of the ACM International Conference on Knowledge Discovery and Data Mining (SIGKDD)*, 236–45. New York: ACM Press.

34. Kalnis, P., N. Mamoulis, and S. Bakiras. 2005. On discovering moving clusters in spatio-temporal data. In *Symposium on Spatial and Temporal Databases (SSTD)*, 364–81. Berlin: Springer-Verlag. http://jjcweb.jjay.cuny.edu/sbakiras/papers/sstd05_mc.pdf.

35. Ester, M., H. P. Kriegel, J. Sander, and X. Xu. 1996. A density-based algorithm for discovering clusters in large spatial databases with noise. In *Proceedings of the ACM International Conference on Knowledge Discovery and Data Mining (SIGKDD)*, 226–31. New York: ACM Press.

36. Chen, Y., and J. M. Patel. 2009. Design and evaluation of trajectory join algorithms. In *Proceedings of the ACM International Symposium on Advances in Geographic Information Systems (GIS)*, 266–75. New York: ACM Press.

37. Brinkhoff, T., H.-P. Kriegel, and B. Seeger. 1993. Efficient processing of spatial joins using r-trees. In *Proceedings of the ACM International Conference on Management of Data (SIGMOD)*, 237–46. New York: ACM Press.

38. Jeong, S. H., N. W. Paton, A. A. Fernandes, and T. Griffiths. 2005. An experimental performance evaluation of spatiotemporal join strategies. *Transactions in GIS* 9 (2): 129–56.

39. Bakalov, P., M. Hadjieleftheriou, and V. J. Tsotras. 2005. Time relaxed spatiotemporal trajectory joins. In *Proceedings of the ACM International Symposium on Advances in Geographic Information Systems*, 182–91. New York: ACM Press.

40. Arumugam, S., and C. Jermaine. 2006. Closest-point-of-approach join for moving object histories. In *Proceedings of the IEEE International Conference on Data Engineering (ICDE)*, 86. Piscataway, NJ: IEEE Press.

41. Zhou, P., D. Zhang, B. Salzberg, G. Cooperman, and G. Kollios. 2005. Close pair queries in moving object databases. In *Proceedings of the ACM International Symposium on Advances in Geographic Information Systems (GIS)*, 2–11. New York: ACM Press.

42. Lee, J., J. Han, and X. Li. 2008. Trajectory outlier detection: A partition-and-detect framework. In *Proceedings of the IEEE 24th International Conference on Data Engineering (ICDE)*, 140–49. Piscataway, NJ: IEEE Press. http://www.cs.uiuc.edu/~hanj/pdf/icde08_jaegil_lee.pdf.

43. Alvares, L. O., V. Bogorny, J. A. F. de Macedo, B. Moelans, and S. Spaccapietra. 2007. Dynamic modeling of trajectory patterns using data mining and reverse engineering. Paper presented at International Conference on Conceptual Modeling (ER2007), Auckland, New Zealand, 149–54. https://doclib.uhasselt.be/dspace/bitstream/1942/7876/2/CRPITV83Alvares-1.pdf.

44. de Vries, G., and M. van Someren. 2008. Unsupervised ship trajectory modeling and prediction using compression and clustering. Paper presented at Belgian-Dutch Conference on Machine Learning, 7–12.

45. Andrienko, G., N. Andrienko, and S. Wrobel. 2007. Visual analytics tools for analysis of movement data. *ACM SIGKDD Explorations* 9 (2) :38–46.

46. de Vries, G. K. D., W. R. van Hage, and M. van Someren. 2010. Comparing vessel trajectories using geographical domain knowledge and alignments. In *IEEE International Conference on Data Mining Workshops (ICDMW)*, 209–16. Piscataway, NJ: IEEE Press.

47. Piciarelli, C., and G. L. Foresti. 2006. On-line trajectory clustering for anomalous events detection. *Pattern Recognition Letters* 27 (15): 1835–42.

48. Bakalov, P., M. Hadjieleftheriou, E. Keogh, and V. J. Tsotras. 2005. Efficient trajectory joins using symbolic representations. In *Proceedings of the 6th International Conference on Mobile Data Management*, 86–93. New York: ACM Press.

49. Alvares, L. O., V. Bogorny, B. Kuijpers, J. A. F. de Macedo, B. Moelans, and A. Vaisman. 2007. A model for enriching trajectories with semantic geographical information. In *Proceedings of the 15th ACM SIGSPATIAL International Symposium on Advances in Geographic Information Systems (GIS)*, 1–8. New York: ACM Press.

50. Gudmundsson, J., J. Katajainen, D. Merrick, C. Ong, and T. Wolle. 2009. Compressing spatio-temporal trajectories. *Computational Geometry* 42 (9): 825–41.

51. Zhu, F. 2011. Mining ship spatial trajectory patterns from AIS database for maritime surveillance. In *Proceedings of the IEEE International Conference on Emergency Management and Management Sciences (ICEMMS)*, 772–75. Piscataway, NJ: IEEE Press.

52. Bäuerle, T., J. Graichen, K. Meyer, S. Seum, M. Kulessa, and M. Oschinski. 2010. *Integration of marine transport into the European emissions trading system: Environmental, economic and legal analysis of different options*. Dessau-Roßlau, Germany: Federal Environment Agency (Umweltbundesamt).

53. United Nations Conference on Trade and Development (UNCTAD). 2008. *Review of Maritime Transport 2008*. New York: United Nations. http://unctad.org/en/Docs/rmt2008_en.pdf.

54. International Maritime Organization (IMO). 2013. AIS transponders. http://www.imo.org/ourwork/safety/navigation/pages/ais.aspx.

55. Knezo, G. J. 2004. Sensitive but unclassified and other federal security controls on scientific and technical information: History and current controversy. Congressional Research Service, Library of Congress. http://www.fas.org/sgp/crs/RL31845.pdf.

56. de Vries, G., and M. van Someren. 2010. Clustering vessel trajectories with alignment kernels under trajectory compression. In *Proceedings of the 2010 European Conference on Machine Learning and Knowledge Discovery in Databases*, 296–311. Berlin: Springer-Verlag.

57. Lane, R. O., D. A. Nevell, S. D. Hayward, and T. W. Beaney. 2010. Maritime anomaly detection and threat assessment. In *Proceedings of the International Conference on Information Fusion*, 1–8. Piscataway, NJ: IEEE Press.

58. Carminati, B., E. Ferrari, J. Cao, and K. L. Tan. 2010. A framework to enforce access control over data streams. *ACM Trans. Information and System Security* 13 (3): 28.

59. Lindner, W., and J. Meier. 2006. Securing the borealis data stream engine. In *Proceedings of 10th International Database Engineering and Applications Symposium*, 137–47. IEEE Computer Society: Washington, DC.

60. Fairplay, Ltd., Lloyds Register of Ships, 8410 N.W. 53rd Terrace, Suite 207, Miami, FL.

61. Kiss, J. 2011. Facebook places deals to target local business ads in UK and Europe. *The Guardian* (online edition), January 31. http://www.guardian.co.uk/technology/2011/jan/31/facebook-places-deals-uk-europe.

12 Above and Below the Ocean Surface: A WSN Framework for Monitoring the Great Barrier Reef

Cesare Alippi and Manuel Roveri

CONTENTS

12.1 INTRODUCTION AND MOTIVATION

Monitoring systems for environmental management applications require real-time data collection and analysis at temporal and spatial scales that have historically been cost prohibitive or beyond the capacity of available technologies. Knowledge growth and technical capacity, however, are rapidly changing globally, and the opportunities offered by modern electronics, which nowadays pervades every aspect of our daily lives, are particularly interesting for research activities that concern the environment around us [1].

Web-based data delivery and satellite communications links are two prime examples. As a result, the collection of environmental data and their analyses to ascertain real-time system status and performance, and the presentation of "ready to use" results, are now major requirements of any new technology. This development is being driven by the need for higher speed, lower price, and improved quality data to be used by government, management, and scientific clients. Similarly, there is a growing need to effectively monitor a wide range of infrastructures and industrial processes so that their sustainability is maximized and their security guaranteed (examples include the water, mining, agriculture, and transport industries). This necessitates an increase in the temporal and spatial resolution of sensors as well as the availability of "smart" monitoring systems that can be tasked by the user in accordance with the dynamics of the environment being monitored and the operational needs of the particular user.

In this context, wireless sensor networks for monitoring marine environments are a fundamental instrument when looking into the depths of our oceans. However, the potential impact of the technological system used is undoubtedly one of the main critical points. The wireless characteristic of these sensor elements makes for an ideal instrument, since they allow for reducing the environmental impact of the monitoring system (e.g., no wires need to be deployed). In the context of environmental monitoring, specifically in the highly avant-garde research sector of water applications, the wireless feature can be envisaged to build up advanced systems dedicated to marine environments, such as coral reefs.

In this direction, the Great Barrier Reef, stretching for over 2000 km along the entire coast of Queensland, Australia, is the largest of tropical coral barriers and is now a World Heritage area safeguarded by UNESCO. This splendid natural resource is one of the oldest and most species-rich ecosystems on Earth: Indeed, some types of coral take hundreds of years to develop into vast colonies on the ocean floor. The immense natural, economic, and social value of the Barrier, together with the particular magnitude of a system of such dimensions, poses no few challenges to which modern technology—both in the field of sensors and of integrated communication between complex devices and ongoing multidisciplinary scientific research—can now give a concrete response. The Great Barrier Reef is an ideal study case for a project concerning the installation of a large environmental monitoring network in a protected area. In addition, the unconventional operational environment represents a further element of innovation and advanced research.

The aim of this chapter is to present an aquatic monitoring framework based on wireless sensor networks (WSNs) that is scalable, adaptive with respect to topological changes in the network, power-aware in its middleware components, and endowed with energy-harvesting mechanisms to increase the lifetime of the monitoring system. The proposed framework addresses all aspects related to environmental monitoring: sensing, local and remote transmission, data storage, and visualization. This interactive network of sensors, data, analysis, and users represents an entirely new way of monitoring and understanding our environment and the anthropogenic impact. An effectively integrated and seamless system is required to meet current and foreseeable research and management needs rather than elements or fragments that require specialist operators or expensive additions to become fully useful.

The proposed WSN-based monitoring system has been deployed at Moreton Bay, Brisbane (AUS), to deliver temperature and luminosity data of the marine ecosystem at different depths. However, this initial sensor configuration can be simply extended by adding application-related sensors and actuators. Data cover scales of time and space, information that has, until now, been unavailable due to cost, technical feasibility, or a combination of the two.

The developed system leads to the following outcomes:

- Ability to monitor the marine environment at multiple scales simultaneously so that it is possible to validate biological models for these ecosystems
- A significant improvement in the prediction of the occurrence of ecological phenomena such as toxic algal blooms and climate-related loss of species and overall biodiversity
- Availability of precious and real time data for the development of risk-based and early warning systems (e.g., hurricanes formation) to underpin management response and the longer-term sustainable management of coastal marine resources

The proposed system uses the marine ecosystem as a bench-test for the technology due to the rigorous demands that equipment faces in marine conditions. Once established, the same system can be adapted and embedded into other existing systems used for industrial and environmental assessment, including process monitoring, risk management, event forecasting, and emergency response. Similarly, the system will be directly applicable to water monitoring across urban, agricultural, and natural systems.

The structure of the chapter is as follows. Section 12.2 critically reviews the principal environmental monitoring applications present in the literature. Section 12.3 presents the network architecture of the proposed monitoring system. Section 12.4 describes the sensor nodes and the gateway. Section 12.5 introduces the energy-harvesting mechanisms and the energy policies to prolong the lifetime of the system. Section 12.6 briefly presents the real-time data storage and visualization system. Finally, Section 12.7 discusses the real deployment of the system at Moreton Bay (Brisbane, AUS) together with an analysis of the problems encountered during the deployment and the acquired data both at the application and sensor levels.

12.2 ENVIRONMENTAL MONITORING APPLICATIONS: CRITICAL ANALYSIS AND REVIEW

A sizeable reduction in the cost of sensors has enabled the rapid deployment of a large quantity over a vast area of *wireless monitoring units* able to measure the relevant environmental parameters and, at the same time, share the information gathered, communicating with their neighbors and, in some cases, starting up a more complex level of data processing within macroareas. The constraints on resources that concern a wireless sensor network, especially in terms of energy consumption, require an integrated collaborative strategy between the different communication levels within the same network. A traditional centralized approach provides for an information exchange flow from each sensor to a single base station, where all the information is gathered and processed to extract the content required for study.

This solution may work well when the number of sensors involved is relatively small and when the transfer information content is not too large. However, when the area to be monitored is particularly vast and the number of sensors used may even reach several millions, a distributed approach may be more appropriate. Radical changes in the design of these kinds of networks necessarily call for the development of new strategies for the management of information processes and for choosing the most appropriate collocation of the computational complexity levels to distribute the various levels of the network itself.

Topological aspects are also of particular importance in guaranteeing an effective elaboration of the signals at different network levels. Undervaluing this aspect can jeopardize the validity of the results obtained; in this respect, the study is now being steered mainly toward an analysis of so-called cluster-based systems. A main node, called a *gateway*, collects information only from a set of sensors under its control and then sends the filtered and compressed data to a base station on land.

An independent capacity to pre-elaborate sensor data constitutes a fundamental requisite for the creation of top-level survey structures. This working mechanism is, in fact, only one module of the network and can potentially be replicated as desired, guaranteeing maximum scalability of the system itself in order to cover far wider areas.

In addition, we must stress that the purely managerial aspect of a large environmental monitoring network is particularly critical because studying these systems necessarily means developing coordination skills among the various actors in a complex communication structure. In fact, a project of this complexity involves both federal and state government agencies, environmental authorities, universities, industries, private individuals, and numerous stakeholders, with an evident need for constant coordination to make the best use of the resources available.

We now survey existing works in the area of WSN-based environmental monitoring to better explain how our work fits in the global context.

Mainwaring et al. [2] presented a star-based topology for monitoring seabird habitats (the gateway collects data from the sensor nodes and forwards them to a remote control station for further processing). A solar-energy-scavenging mechanism has been envisaged only at the gateway level, leaving the sensor nodes battery powered. Since the communication refers to the sensing unit-gateway and gateway-base station segments, the gateway must always be on. No adaptation schemes have been reported.

Werner-Allen et al. [3] proposed a system for monitoring volcanic eruptions; no energy-harvesting solutions have been included for the gateway and the three sensor units. Although the transmission protocol is based on a traditional scheme involving packet retransmission, a large number of packets were lost due to weather conditions and the presence of obstacles.

A more complex WSN architecture was proposed by Hartung et al. [4] involving the use of multihop WSNs for monitoring wildland fires. All units are battery powered without energy-harvesting mechanisms. The limited adaptation ability of the system requires human intervention at the software level for addition of new nodes.

Biagoni and Bridges [5] have presented a WSN for studying rare and endangered species of plants in the Hawaii Volcanoes National Park. Units, equipped with cameras among other sensors, are battery powered; no energy-harvesting mechanisms are envisaged. A multihop protocol, which implements a synchronization mechanism and an adaptive routing strategy, has been considered

Zhang et al. [6] have proposed a WSN for tracking zebras. Each unit is equipped with a GPS (Global Positioning System) sensor. Acquired data are periodically collected by a data mule installed on a manned mobile base station. Unfortunately, such an approach is not effective for applications requiring a large WSN or not allowing a data mule. Solar panels recharge the units' batteries.

A WSN designed to detect the cane toad in northern Australia was presented by Hu et al. [7]. The WSN units acquire data through acoustic sensors, locally process them, and then transmit the data to a gateway that, in turn, sends the frog "presence/absence" information. Neither energy-harvesting nor power-aware solutions nor sophisticated energy-management policies have been reported.

He et al. [8] worked on one of the major efforts in the WSN field. The application refers to the development of a WSN to identify mobile targets within a surveillance application. The 70 units also manage the occurrence of failures by periodically rebuilding the routing paths. No energy-harvesting solutions have been considered; conversely, an effective duty-cycle mechanism is envisaged for keeping power consumption under control.

Baggio [9] presented a system to monitor a crop field with temperature and humidity sensors. The application is based on a low-duty cycle and implements a multihop protocol to forward data from sensor nodes to a gateway, which is connected by a WiFi link to a base station. Routing tables are periodically rebuilt to provide robustness; some sensorless units are deployed and act as communication relays.

Wark et al. [10] considered a system to measure soil moisture and monitor grass conditions; units are equipped with solar panels. WSNs are also used to study cattle behavior, as shown in the work of Sikka, Corke, and Overs [11] and Kansal and Srivastava [12], adopting units equipped with GPS and solar panels to track animals' movements. All data are sent to a gateway placed in the paddock.

In all of these deployments, wherever adopted, the considered energy-harvesting mechanism is based on an on-off charging scheme. While this is effective in optimal sun conditions, its efficiency drastically falls (the charging does not occur) when there is insufficient radiation available (e.g., in presence of a partly cloudy sky, mist, morning and late hours, dust on the solar panel, etc.). In contrast, for effective solar-energy harvesting in adaptive radiating situations, a Maximum Power Point Tracker (MPPT) scheme must be addressed to maximize charging efficiency, as the MPPT adapts on-line the solar cell working point with the solar radiation.

At the same time, an energy-harvesting mechanism impacts on the network topology and, in turn, on the communication protocol. In fact, units run out of energy when no power is available (or when the batteries are exhausted), and they need to be switched off and then switched on once the energy is back. This start/restart mechanism needs both hardware and software modifications to be an effective and a robust power-aware communication protocol.

It immediately arises that a credible deployment requires units and gateways to be equipped with MPPT-based energy harvesting mechanisms as well as a communication protocol robust with respect to perturbations affecting the Quality of service (QoS) and adaptive with respect to changes in topology.

12.3 DESIGNING THE SYSTEM: THE NETWORK ARCHITECTURE

In its simplest architecture, the proposed WSN, which has been presented by Alippi et al. [13], is characterized by a cluster topology ruled by a cluster head (here acting as a gateway); the cluster constitutes the core of a more sophisticated hierarchical architecture obtained by adding other clusters to the network (see Figure 12.1).

Gateways forward the collected information to the base station. A multihop approach at the gateway level would constitute a different option for conveying data to the base station that was not requested in the considered application by marine biologists.

For communication at the cluster level and the gateway, one may use the same transmission protocols (e.g., through a Frequency Division Multiple Access—FDMA—acting at the gateway communication level).

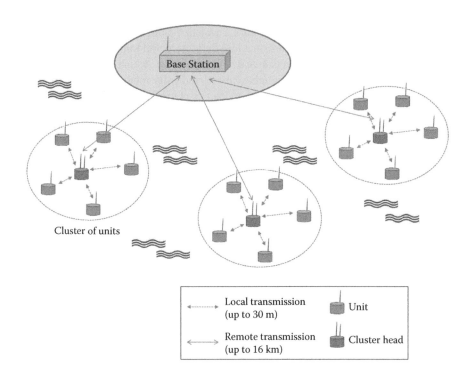

FIGURE 12.1 Network architecture.

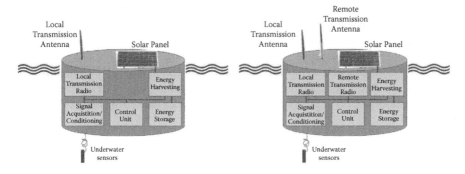

FIGURE 12.2 (a) Sensor node architecture; (b) gateway architecture.

At the cluster level, a traditional hierarchical Time Division Multiple Access (TDMA)-based solution would be particularly appealing for simplicity and energy efficiency, but surely it is not adequate in a situation where the network topology is subject to a continuous change, and it might also suffer from intracluster frequency interference. While the latter issue can be solved by considering the frequency-allocation mechanism proposed by Aardal et al. [14], the former requires design of an ad hoc TDMA-based protocol able to deal with adaptation in the network topology while also keeping in mind energy-savings aspects. Such a protocol has been presented by Alippi et al. [13].

Each sensor node (see Figure 12.2a) is composed of five main modules: control unit and data processing, signal acquisition and conditioning, local transmission radio, energy-harvesting mechanisms, and energy storage. The gateway (see Figure 12.2b) is a sensor-node unit augmented with a long-range communication ability that allows the remote transmission to the base station.

Nodes and gateways, fully designed in terms of both hardware and software by our group, are inserted in waterproof buoys (see Figure 12.3); the underwater sensors (developed at University of Queensland) live outside the buoys.

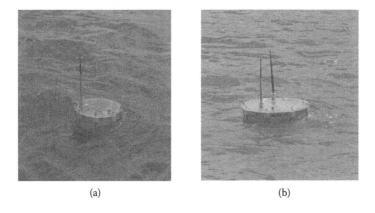

(a)   (b)

FIGURE 12.3 (a) A sensor node; (b) the gateway.

Energy is provided by two 0.5-W and eight 0.5-W polycrystalline solar cells for nodes and gateways, respectively, with a tandem battery solution as a storage means. This setup is described in Section 12.5.

One dipole omnidirectional antenna is present on the nodes for communicating to the gateway, which itself mounts two omnidirectional antennas: one to communicate with the local cluster network and the other for establishing a radio link to the ground control station. Details regarding both the sensor nodes and the gateways are given in the next section.

12.4   DESIGNING THE SYSTEM: SENSOR NODES AND GATEWAYS

Nodes and gateways have been designed to be interchangeable, with the unique hardware difference being the presence of the radio link module for the gateway.

A buoy's electronics is composed of two circular boards stacked vertically and linked through connectors. The upper board, which is given in Figure 12.4, contains underwater sensor signal conditioning, a long-range radio transceiver (for the gateways), a short-range radio transceiver, and the unit's main control CPU. The lower board, which will be detailed in the next section, is responsible for the unit's power management and storage; the upper board is responsible for managing processing, signal conditioning, and radio communication.

The signal acquisition and conditioning module of the upper board is composed of a PAR (Photosynthetic Active Radiation) signal conditioning, a TH (thermal) signal conditioning, and a MS (moisture) signal conditioning. The device deals with analog acquisition of the physical quantities to be monitored (e.g., temperature or brightness

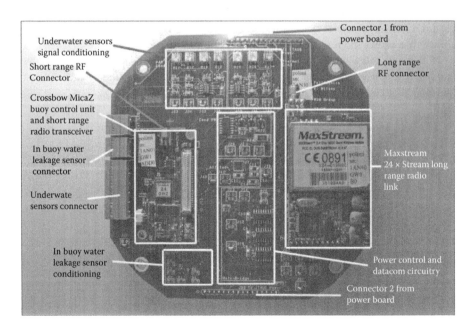

FIGURE 12.4   [Upper board] Signal board main modules.

as well as humidity sensors) and the subsequent signal-conditioning phase. Ad hoc solar radiation and temperature sensors have been developed by the University of Queensland for this application, since effective off-the-shelf underwater sensors were not available on the market. The module provides an electronic interface commanded by the control unit; it performs traditional signal filtering before an analog-to-digital conversion, and it allows underwater sensors to be turned on and off for energy awareness. Moisture sensors have been employed to detect possible leaks in buoy-sealing o-rings or in defective sealed connectors, and these sensors send an alarm in the event of an electronics failure or buoy sinking. A moisture-sensor signal was also placed in the underwater sensor housing for the same reasons.

The main control unit is a Crossbow MPR2400 (MicaZ) unit [15], which has a 16-bit microcontroller and acts as the main CPU. We considered MicaZ for their large market availability and academic usage, the possibility of using the open-source TinyOS operating system, and the contained power consumption of the unit.

As regards the local and remote transmission radios we used in the proposed framework, the Chipcon CC2420 (MicaZ) low-range radio module for cluster communication allowed us to easily cover a 2800-m^2 area (30-m radius). However, nowadays we would have opted differently for radio modules able to fully support the Zigbee communication protocol [16], for example modules provided by Jennic Wireless Microcontroller [17] or Freescale [18].

The long-range transceiver, which is present in the upper board of the gateway, is a MaxStream 2.4-GHz Xstream Radio Modem [19], able to provide 50 mW @ 2.4 GHz of RF power at the long-range antenna. Since power consumption of this module is particularly high (being around 0.75 W in transmission), the main control unit switches it on and off at fixed intervals of time to communicate with the ground station. When it is not in the transmission phase, the radio is posed in a low-power sleep state, and its DC/DC power supply is switched to burst mode control to further save energy, as described in the next section. Data to be broadcast are provided to the radio modem by the main control CPU through its serial communication line.

12.5 DESIGNING THE SYSTEM: ENERGY HARVESTING AND STORAGE

The lower board contains solar cell control circuitry, battery control circuitry, a high-efficiency DC/DC power supply module, and an energy management CPU that controls the power flows. The printed circuit board is given in Figure 12.5, where subunits have been outlined.

Energy generation, storage, data acquisition, DC regulation, and cold-starting activities are coordinated by a dedicated 8-bit microcontroller that operates independently from other subsystems. The microcontroller executes a relatively simple embedded C code at the extremely low-power-consumption 32-kHz clock frequency. In addition, the microcontroller acts as a hard-failure watchdog for higher-level subsystems by simply toggling down their power supply once failure conditions are detected.

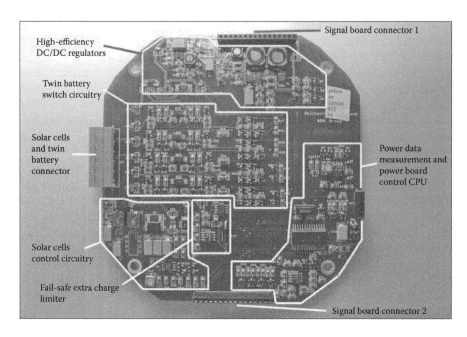

Signal board connector 1

High-efficiency
DC/DC regulators

Twin battery
switch circuitry

Solar cells
and twin
battery
connector

Power data
measurement and
power board
control CPU

Solar cells
control circuitry

Fail-safe extra charge
limiter

Signal board connector 2

FIGURE 12.5   [Lower board] Power board and main subunits.

The energy-harvesting module is composed of a CPU and a controlled step-up DC/DC converter based on the Maximum Power Point Tracker (MPPT) circuit suggested in the literature [16]. The MPPT optimally harvests solar energy by adapting the working point of the solar cell to maximize energy transfer from the cells to the batteries. The system supersedes existing energy-harvesting methods by harvesting energy even when the cell is not directly exposed to the optimal radiation or when the solar radiation is low, as can happen in outdoor applications when the panel surface becomes dusty or is covered with water and/or marine salt, or when clouds change the intensity of the solar radiation.

The power extracted by the MPPT circuit is then directed to the storage mean, here designed as a special twin battery system. The basic idea behind this energy-storage solution is to separate the charge and the discharge phases of the batteries, i.e., the system allows the batteries to be charged and discharged at separate intervals of time. With this solution, we solve one of the major drawbacks in small solar-powered devices, namely, the partial charge/discharge suffered by batteries during the day/night cycle: During the day, solar power is conveyed to batteries (charge phase), while during the night no solar power is present and batteries provide power to the sensor units (discharge phase). It is a common practice (as we did) to employ batteries with energy capacity larger than the solar cell's daily energy production and daily system energy consumption in order to be able to store energy and permit the system to supply power even in the case of bad weather. For this reason, in normal weather conditions, batteries are partially charged and discharged every day.

Unfortunately, chemical batteries cannot withstand a prolonged set of partial charge/discharge cycles, since many secondary chemical effects arise: Forcing a battery into a prolonged cycle of partial charge/discharge causes a severe reduction of battery nominal energy capacity and, as a consequence, severely limits its performance. (This effect is often referred to as the *memory effect* of chemical batteries.) One way to solve this unwanted phenomenon is to separately perform full discharge and full charge cycles.

We employed two identical battery packs and implemented a very straightforward idea: While one battery pack powers the system (and thus is in a discharge mode), the other is under charge. Once the former is discharged, an embedded circuit allows the battery packs to be inverted. In this way, one battery pack is always in charge while the other is always being discharged.

A special ad hoc battery-switching unit has been designed and developed. The unit is able to connect the two battery packs either to the output of the MPPT converter or the input of the DC/DC regulators in response to dedicated commands coming from the control CPU. Moreover, the unit is able to completely disconnect both battery packs from the output of the MPPT circuit and/or from the input of DC/DC regulator whenever the energy-control CPU issues such a command. This point is relevant, since units need to be disconnected from batteries for lack of energy (hence allowing them to be recharged provided the availability of energy) and then reconnected once sufficient energy is back. In particular, to prolong battery lifetime, the control CPU completely disconnects battery packs from units when both are in undercharge conditions (i.e., in the case of a prolonged lack of solar power and the batteries are exhausted). On the contrary, if an overcharge condition occurs on a battery pack under charge, the battery is disconnected from the MPPT circuit and posed in an idle state (the system is powered by the other battery pack) while generated power is automatically redirected to a special wasting unit.

In commercial WSN units, lack of energy coincides with switching them off; to the best of our knowledge, no mechanisms are available to grant a new restart of the unit when energy is available. Since this is a main issue to grant a credible deployment, we developed a nontrivial circuit and procedures to grant a cold restart for the units when both battery packs are in a deep discharge state. By solely relying on solar cells (i.e., without any external maintenance intervention), the circuit acquires energy, stores it in a battery, and waits to have enough energy before powering the unit modules.

The last module of the board is composed of DC/DC regulators whose duty is to power the system. We employed four NiMH-cell battery packs with 4.8 V nominal voltage (different unit modules would require different voltages for powering). In particular, we generated three DC powering buses: 3.3 V for the main control unit, 5 V for the radio link, and 7.5 V for the underwater sensors developed at the University of Queensland.

12.6 DESIGNING THE SYSTEM: DATA STORAGE AND PRESENTATION

Measurements acquired by units and transmitted by the gateway finally reach the control center for data storage and aggregation (see Figure 12.6). In particular, measurements and their time stamps are stored in a database based on a multithreaded,

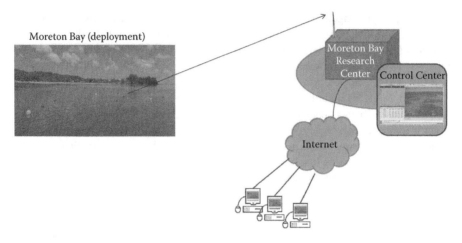

Moreton Bay (deployment)

Moreton Bay
Research
Center   Control Center

Internet

FIGURE 12.6 An overview of the system.

multiuser SQL DBMS MySQL Server 5.0 running on a Linux OS. In addition to the application measurements, we also stored the status of the network, defined as the "connected, not connected, not registered" label for each unit, the voltage of the batteries, the input power coming from the solar panel, the humidity level, and the possible presence of water within the buoy.

Operators easily query the database at the control station or remotely, through the Internet, thanks to a proprietary software application (see Figure 12.7 for two views of the system). In particular, users may inquire about temperature, brightness, and status of the batteries for a specific node or for all connected nodes by specifying the time interval of interest. Moreover, an immediate graphical user interface provides the state of all the nodes in the network and, for each connected node, the most

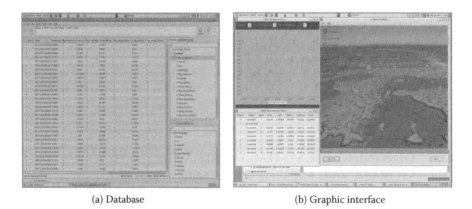

(a) Database   (b) Graphic interface

FIGURE 12.7 Database and graphical interface at the base station.

FIGURE 12.8   Final deployment at Moreton Bay (Brisbane, Australia).

recent measurement acquisitions (temperature and brightness) and information about its state (status of the batteries, solar power absorption, and alarms) both numerically and graphically.

12.7   DEPLOYMENT AT MORETON BAY (AUSTRALIA)

After a careful assessment of the environmental context in which the monitoring project was going to be set up, and of the environmental impact that the installation of such a system could determine in the area under study, Moreton Bay (see Figure 12.8), located on the southeast coast of Queensland, was chosen as the location for the pilot project. The University of Queensland already had a research station here, but its monitoring systems were obsolete and rather costly. The outcome of the project is thus to build up the first coral monitoring network in this site: Thousands of measurements taken during the day are able to constantly monitor the marine environment in which the coral, one of the planet's most delicate living organisms, is now struggling to survive. The need for a detailed study of phenomena occurring in this natural habitat and the limitations of traditional survey systems have encouraged research and the application of a highly innovative solution.

The developed WSN-based monitoring system deployed at Moreton Bay is composed of nine units (immediately scalable up to 70 in a plug-and-play fashion) and a gateway; devices are inserted in buoys anchored to the coral reef (see Figure 12.9 for a top inside view of a unit). The distance between gateway and base station was about 1 km; units were deployed in a pseudo-linear configuration as seen in Figure 12.8.

FIGURE 12.9 The WSN unit.

The sampling frequency of the temperature and the brightness sensors is 1 Hz; acquired measurements are averaged (to reduce acquisition noise) and sent to the gateway every 30 s. Currently, the result is a set of buoys, approximately 24 centimeters in diameter and about half as much in thickness (see Figure 12.9), that transmit information picked up from the environment in real time, thanks to high-sensitivity wireless sensors. In this way it is possible to appreciate even minimal variations in the environmental parameters that might influence the life of living organisms inhabiting the Great Barrier Reef.

The number of measurement points in Moreton Bay will soon grow into a capillary network monitoring every minimal environmental variation. The innovative buoys, conceived by a group of engineers and researchers from the Politecnico di Milano, are above all the fruit of research carried out over many years in close contact with a group of biologists from the Centre of Marine Studies at the University of Queensland.

In addition, the fact that data is transmitted in real time overcomes the enormous limits of the old monitoring system, which required a boat and divers to place the sensors, who then returned to collect them after several weeks. Quite apart from the substantial cost of sending divers by boat (at least twice per monitoring mission), the old system did not allow for an immediate study of the phenomena observed. Moreover, only after a lengthy period was it possible to identify any breakdown in the sensors that may have jeopardized the entire mission, with no possibility of intervention or real-time control.

The current sensors measure temperature, salinity, and luminosity, and are able to send information automatically to a buoy working as a data collector (i.e., the gateway), which relays it to the observation station on land. This mechanism enables a more efficient coverage than previously, with more measurement points and a higher sampling frequency, improving the quality of monitoring and the biologists' capacity to create predictive models of phenomena affecting the corals.

In this perspective, a natural further application for the immediate future is the prevention of environmental disasters, such as the loss of crude oil from an oil tanker or shipping accident (events that have unfortunately not been so uncommon in recent years).

Finally, the whole system has been designed to be of low environmental impact: In fact, the buoys use solar energy to recharge their batteries and become totally self-sufficient as measurement points (see Figure 12.9). This solution also has undeniable advantages from an economic point of view. The production cost of a buoy, which is now around 300 Euro, is destined to fall drastically. Thus the same technology as that tested in Australia could find important applications in developing countries, for completely different purposes, enabling the monitoring of phenomena barely imaginable today.

12.7.1   CRITICAL ASPECTS OF THE DEPLOYMENT

In the case of marine monitoring, the buoy is one of the most important issues, since watertightness must be guaranteed. In turn, this implies that physical connectors between the electronic boards, which remain inside the package, and external elements (i.e., sensors and antenna) must be watertight. To further reduce the water infiltration risk, magnetic switches have been considered to activate the electronics: Units can be enabled/disabled without any physical connector with the buoy inside. Fixing elements such as screws, hooks, cables, and the anchorage bolt must be made with special steels able to resist corrosion in the marine waters.

We included some utility sensors inside the buoy, such as humidity and water sensors, to detect as quickly as possible the formation of water or water infiltrations. (The sensors allowed us to detect water infiltration in the gateway during the first deployment.) Temperature within the buoy was controlled by a thermal exchanger, a metallic steel closure placed on the bottom of the buoy in direct contact with the sea.

A further critical aspect was the anchorage system and its interaction with waves during high and low tides. Buoys have been anchored with a mooring cable to a reinforced concrete block at the sea bottom. Waves induce severe flutters on the buoys that impact on the mooring cable (which could break, causing the loss of the buoys) or affect the sensors. Moreover, when the distance between adjacent buoys is not adequate, the mooring cables may get twisted, causing possible tears. For this reason, we used elastic mooring cables able to absorb the strong swinging the waves can induce on the buoys. Moreover, we used a specific hook (between the buoy and the mooring cable) to maintain the buoy in a horizontal position, even in the case of strong streams.

There is another problem that affects the measurements: the formation of algae on the sensors, which affects the data-acquisition process. Calibration can be envisaged to reduce this loss in accuracy effect, but a periodic cleaning of the sensor (e.g., once per year) is required.

As a final issue, the deployment of units needs to be adequately signaled with signaling buoys, and permission for the deployment needs to be released by the relevant authority.

12.7.2   CURRENT RESULTS AND DISCUSSIONS

In this section, we present and discuss the measurements acquired by a node in a four-day acquisition campaign (from 2007/11/18 20:06:03 to 2007/11/22 09:20:27). In particular, Figure 12.10 presents the temperature, the brightness, and the solar power generated by the solar panels, while Figure 12.11 shows the state of the batteries.

The behavior of the brightness and the solar power follows the classic day/night cycle. As expected, both brightness and incoming solar power assume a zero value during the night (e.g., samples in the [0, 960] interval for day one), then they rise in the morning (interval [960, 1260]) and afternoon (interval [1800, 2100]). Even the temperature follows the day/night seasonality, with dynamics depending on the thermal inertia of the water.

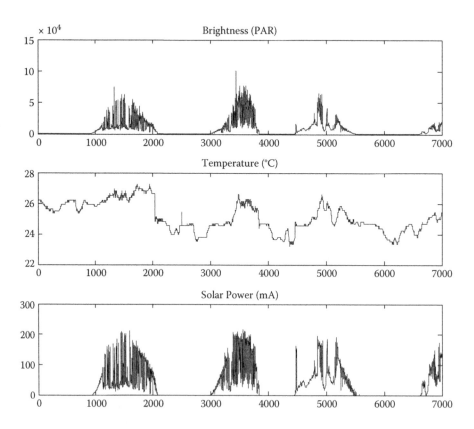

FIGURE 12.10   Node 1: Brightness, temperature, and solar power w.r.t. time (7000 samples acquired over four days.

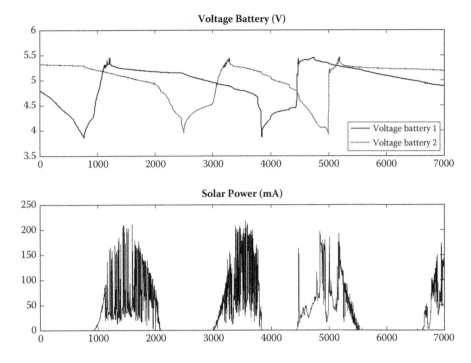

FIGURE 12.11 Node 1: Voltage batteries and solar power w.r.t. time (7000 samples acquired over four days).

Figure 12.11 shows the state of the batteries and the solar power during the acquisition period. At the beginning of the experiment, Battery 1 is active; its voltage decreases the w.r.t. time up to sample 900. Battery 2, in the same period, suffers from self-discharge phenomena, causing a reduction of the voltage even when the battery is not used.

When the voltage of the Battery 1 decreases below 4 V (at sample 900), the energy-harvesting mechanism's module switches between the two batteries and activates Battery 2 (and Battery 1 goes under charge). Between sample 900 and 1200, Battery 1 is recharged by solar energy. When the battery is fully charged (the voltage exceeds 5.5 V), it is disconnected from the energy-harvesting module to prevent overcharge (sample 1200). When a battery is inactive, it suffers from self-discharge phenomena, as can be seen between samples 1200 and 2600. Battery 2 is used up to sample 2600; afterward, the batteries switch: Battery 1 becomes operational and Battery 2 goes under charge.

12.8 CONCLUSIONS AND FUTURE TRENDS

In this chapter, we presented a WSN-based framework for monitoring a marine environment. All aspects of the environmental monitoring system—such as sensing activity, local transmission (from sensor nodes to gateways), remote transmission (from the gateway to the control center), data storage, and visualization—have been designed and implemented. The approach described here differs from other environmental monitoring systems proposed in the literature in that each unit of the WSN is endowed with adaptive solar energy harvesting mechanisms and tandem batteries for optimizing energy storage and prolonging battery lifetime. Finally, the proposed framework has been deployed with success at Moreton Bay, Brisbane (AUS), to monitor the water conditions of a segment of the Australian Coral Reef.

REFERENCES

1. Alippi, C. 2010. Australia: Above and below the ocean surface. The International Gateway. Projects and Partnerships Edizioni Olivares, Milan, Italy.
2. Mainwaring, A., D. Culler, J. Polastre, R. Szewczyk, and J. Anderson. 2002. Wireless sensor networks for habitat monitoring. In *Proceedings of International Workshop on Wireless Sensor Networks and Applications (WSNA)*, 88–97. New York: ACM Press.
3. Werner-Allen, G., J. Johnson, M. Ruiz, J. Lees, and M. Welsh. 2005. Monitoring volcanic eruptions with a wireless sensor network. In *Proceedings of Second European Workshop on Wireless Sensor Networks*, 108–20. Piscataway, NJ: IEEE Press.
4. Hartung, C., R. Han, C. Seielstad, and S. Holbrook. 2006. FireWxNet: A multi-tiered portable wireless system for monitoring weather conditions in wildland fire environments. In *Proceedings of International Conference on Mobile Systems, Applications and Services*, 28–41. New York: ACM Press.
5. Biagioni, E., and K. Bridges. 2002. The application of remote sensor technology to assist the recovery of rare and endangered species. *International Journal of High Performance Computing Applications* 16 (3): 315–24.
6. Zhang, P., C. M. Sadler, S. A. Lyon, and M. Martonosi. 2004. Hardware design experiences in ZebraNet. In *Proceedings of the 2nd International Conference on Embedded Networked Sensor Systems*, 227–38. New York: ACM Press..
7. Hu, W., N. Bulusu, C. T. Chou, S. Jha, and A. Taylor. 2005. The design and evaluation of a hybrid sensor network for cane toad monitoring. In *Proceedings of the 4th International Symposium on Information Processing in Sensor Networks (IPSN)*, 503–8. Piscataway, NJ: IEEE Press.
8. He, T., S. Krishnamurthy, L. Luo, T. Yan, L. Gu, R. Stoleru, G. Zhou, et al. 2006. VigilNet: An integrated sensor network system for energy-efficient surveillance. *ACM Trans. on Sensor Networks (TOSN)* 2 (1): 1–38.
9. Baggio, A. 2005. Wireless sensor networks in precision agriculture. Paper presented at ACM Workshop on Real-World Wireless Sensor Networks (REALWSN 2005), Stockholm, Sweden.
10. Wark, T., P. Corke, P. Sikka, L. Klingbeil, Y. Guo, C. Crossman, P. Valencia, D. Swain, and G. Bishop-Hurley. 2007. Transforming agriculture through pervasive wireless sensor networks. *IEEE Pervasive Computing* 6 (2): 50–57.
11. Sikka, P., P. Corke, and L. Overs. 2004. Wireless sensor devices for animal tracking and control. In *Proceedings of 29th Conference on Local Computer Networks*, 446–54. Piscataway, NJ: IEEE Press.
12. Kansal, A., and M. Srivastava. 2003. An environmental energy harvesting framework for sensor networks. In *Proceedings of International Symposium on Low Power Electronics and Design (ISLPED)*, 481–86. New York13. Alippi, C., R. Camplani, C. Galperti, and M. Roveri. 2011. A robust, adaptive, solar powered WSN framework for aquatic environmental monitoring. *Sensors Journal, IEEE* 11 (1): 45–55.

14. Aardal, K., S. van Hoesel, A. Koster, C. Mannino, and A. Sassano. 2007. Models and solution techniques for frequency assignment problems. *Ann. Oper. Res.* 153:79–129.
15. Xbow. 2008. http://www.xbow.com/.
16. Zigbee. 2003. http://www.zigbee.org.
17. Jennic Wireless Microcontroller. 2008. http://www.jennic.com/.
18. Freescale Semiconductor Inc. 2008. http://www.freescale.com/.
19. Maxstream. 2008. http://www.maxstream.com/.

Index

Printed and bound by CPI Group (UK) Ltd, Croydon, CR0 4YY

18/10/2024

01776269-0004